The Musical Engineer

A Music Enthusiast's Guide to Engineering and Technology Careers

CELESTE BAINE

Engineering Education Service Center
Springfield, OR

The Musical Engineer

A Music Enthusiast's Guide to Engineering and Technology Careers

by Celeste Baine

Published by:

Engineering Education Service Center (an imprint of Bonamy Publishing)
1004 5th St
Springfield, OR 97477 U.S.A.
(541) 988-1005
www.engineeringedu.com

Printed in the United States of America

CIP Pending

ISBN 10: 0-9711613-7-2 (pbk.)
ISBN 13: 978-0-9711613-7-5 (pbk.)

How to Order:

Single copies may be ordered from the Engineering Education Service Center, 1004 5th Street, Springfield, OR 97477; telephone (541) 988-1005; Web site: www.engineeringedu.com. Quantity discounts are also available. On your letterhead, include information about the intended use of the books and the number of books you wish to purchase.

Disclaimer

Although the author and publisher have attempted to research all sources exhaustively to ensure the accuracy and completeness of information on the subject matter, the author and publisher assume no responsibility for errors, inaccuracies, omissions, or any other inconsistencies herein.

The purpose of this book is to complement and supplement other texts. You are urged to read all the available literature, learn as much as you can about the field of engineering, and adapt the information to your particular needs.

If you do not wish to be bound by the above statements, you may return this book to the publisher for a full refund.

Acknowledgments

First of all, I'd like to extend a huge thank you to the audio, engineering and engineering technology communities that encouraged my vision and lended a hand in the creation of this book. It is the collaboration and support of these groups that helped me meet the incredibly talented engineers, technicians, consultants, and musicians that contributed to this book. Thanks also to Beau Eastlund, Mark French, Tom McGlew, Ken Pohlmann, and Rick Wyfells for reviewing the book under such a tight deadline and for their industry feedback. Thank you Kelly Eastlund for making my grammar more appealing and palatable. My gratitude also goes to Michael Howes, Michael Godfrey, Keith Hatschek, Gino Sigismondi, Mark Amundson, and Robert Smith for their time and contribution.

Thanks to the Audio Engineering Society for their interest in this project and for the complementary admission to their annual conference.

I am especially grateful to John L. Rice for staying up all night trying to get just the right photos for the book under a tight deadline.

And a big thank you goes to my family for putting up with my obsessions and deadlines.

Credits:
Cover Design: Amy Siddon
Copy Editor: Kelly Eastlund
Photos: John L. Rice and Celeste Baine
Inquiries regarding photographs taken by John L. Rice can be directed to: photos@ImJohn.com, or 2522 N Proctor St. #164, Tacoma, WA 98406. Web: www.ImJohn.com/photos/

Contents

Chapter 4 - Instrument Design and Manufacturing

Chapter 5 - Computer Science/Software Engineering

Chapter 6 - Digital Music

Chapter 7 - Other Careers

Chapter 8 - Getting Started

Appendix

This book is dedicated to all the drummers that were never allowed to bang freely for fear of angering the neighbors.

"Master your simple tools. Make your music low baggage, high mileage; passion before commerce and intelligence before waste." – Brian Eno, legendary musician and producer

About this Book

When I was a kid, music engineers were the people turning the knobs in a recording studio or moving the vertical sliders in a concert. We looked up to them as being the gurus of sound management. They possessed an ear for music that was only developed through serious training or listening experiences. We needed them to make our experience complete. Now that I'm older, in addition to those engineers that were put on pedestals and worshiped as a part of my youth, I've learned that there are many ways for the musically inclined engineer to work in the music business. If you love music and worry that the path to being a professional musician is too full of holes and hills, there are other ways for you to be involved. This book is about much more than just being a traditional music engineer. This book is a gateway. It is designed to open your mind to the wonders of engineering and technology. You will learn scores of things you can do with an engineering or technology degree and an interest in music. You will discover opportunities you hadn't thought about, and you may even begin to look at engineering in a new way. Be prepared to get excited and to learn that technical skills and capabilities can give you a new and different way of seeing the world.

Written for middle school, high school and pre-engineering college students, this book compiles resources, information and stories of engineers who work passionately in the music industry to design new and improved products for music enjoyment. Motivation to develop your skills, imagination to see the finished product, and the energy to see it through are all outstanding attributes of future engineers. In this book, the word engineer is used to mean those with a four-year (or more) engineering degree as well as those with a two-year engineering technology degree (music, audio, electronics, etc.)

Ranging from the design and construction of stadium and studio sound systems, to the design and manufacture of iPods, electronic instruments, gaming sound, MIDI programming, and much more, you will understand what you need to know to work in this industry and find a satisfying and rewarding job as a music engineer. This book presents possibilities you might not have expected. You will see what types of engineers stream live concerts on the Web and create your favorite music software. You'll also find out how to identify companies that will hire you as an engineer. From the software engineers who design ring tones, to the electrical engineers who work on new microphones and speakers, to the computer engineers who create new thrill ride or animatronic sounds, the sound industry has a place for many types of engineers. You will learn what it takes to design music applications and get advice from engineers about how to succeed in the industry. A strong motivation for writing this book is to help you see that engineering can be fun. It's exciting to be on the cutting edge of technology and to make the world a better place to live.

A vital component of engineering success, especially in the music engineering industry, is excellent social and communication skills. If you have ever played in a band, you understand that teamwork is integral to the success of the band. Each player brings different strengths to the band, without which the band can't function as efficiently or achieve the sound it wants. Engineering design works in the same way. Each member of the team contributes, according to their individual strengths, and, as a result, society gets to discover new bands, listen to music in new and better ways, and have more fun.

Because the music industry is so large, we can't cover every aspect of it in this book. Some industries may be left out entirely and some sections may leave you craving more information. If you find yourself in the craving-more-information category, use that energy to your benefit. Contact studios or audio equipment manufacturers to inquire about summer work or co-op opportunities. Go to the Audio Engineering Society meetings. Tour facilities and begin talking to engineers for tips on getting through school and for gathering information about what they do now. Read books about famous recording engineers and how they got their start. Use a program like Garage Band to mix your own music. Take music or voice lessons. Play Guitar Hero on your Playstation®. When you become competent in the program, picture yourself as the software designer. What would you do differently? What extra bells and whistles would you add to the program? Tinker—collect broken digital music players and see if you can fix them, borrow gear from your friends to experiment, trade gear on Craigslist, or volunteer to record your friend's band. Then, you'll truly begin to think and work like an engineer. Don't be afraid to fail a few times, either.

Thomas Edison is one of the most prolific inventors of all time. What made Edison so great was that he believed that every failure brought him closer to success. As a result of a lifetime of work and tens of thousands of failures, he held over 1,000 patents for his successful inventions. Failure is a rite of passage to success.

Some of the most amazing inventions and technologies on the market today exist because one engineer had an idea. Look back at old pictures of the Vitrola. It had a hand crank, but people wanted it to play without cranking. Change came about because of engineering. Year after year, engineers returned to the drawing board and made the Vitrola better. Eventually, it evolved into the record player that evolved to a CD player that is now a digital music player. What will hearing music look like in another 10 years? It's up to you and your imagination to find out.

If you want an exciting and diverse career, an engineering degree can blow open the doors of opportunity. But don't be misled, this path is by no means easy. If you don't love it, it will be even harder. It requires long hours (both in school and out of school), the work is challenging, you have to be very organized, good with people, and able to work at a crazy pace.

Are you ready? If so, strap on your air guitar and let's get going....

CHAPTER ONE

For the Love of Music

Music is pleasurable. Music is beautiful, relaxing, and fun. It develops the senses, teaches rhythm, and helps develop language and social skills. Music stimulates creativity, reduces stress, and improves your mood. According to the American Music Conference (AMC), music is a major stress reducer that increases melatonin, the hormone that increases the immune system's natural cancer-killing cells.

The AMC goes on to say that, "Young kids who make music show improved spatial-temporal reasoning, which is the foundation of later success in math and science." Reports from the Silicon Valley show that an abundance of scientists and engineers are also practicing musicians. In addition, active listening (playing the air-guitar or singing along to your favorite tune) is known to improve brain power. Students who participate in music get better grades than those who don't.

The value of hearing and listening

Before we really get started, we need to talk about the importance of hearing protection in the music production business. There is a vital need to protect and take care your hearing so you can keep it. Noise-induced hearing damage comes from the destruction of the tiny hair cells (cilia) in the inner ear. At loud noise levels, those delicate hair cells lie flat and are destroyed. When the exposure is frequent or severe enough, the hair cells stay flat permanently and as more and more of the hairs are destroyed you will lose more and more of your hearing. Unfortunately, damaged hair cells in your ears cannot be repaired or replaced.

Excellent listening skills are also an asset in all aspects of life. Critical listening is probably one of the most important classes for audio and music engineers and technologists. Discerning between echoes and reverb or between +3 decibel (dB) at 1 kilohertz (kHz) and +6 dB at 5 kHz is essential. Can you tell the difference between preamp overload and speaker distortion? Can you distinguish between an oboe and a clarinet, and between a guitar in tune and one that's not? These are diamond skills—and they are some of the best skills that you can offer a future employer.

The world of engineering and engineering technology

Choosing between engineering and engineering technology (ET) is a matter of preference for hands-on vs. theoretical work as well as consideration of working environment, opportunities for advancement, financial constraints, and time in school. An engineer and a technologist both go to school for four to five years. A technician usually goes to school for two years to obtain an associate's degree or gets a certificate in one year or less.

Engineering technologists and technicians use the principles and theories of science, engineering, and mathematics to solve technical problems in research and development, manufacturing, sales, construction, inspection, and maintenance. Their work is more limited in scope and application-oriented than that of scientists and engineers. Many engineering technicians assist engineers and scientists, especially in research and development. Others work in quality control, inspecting products and processes, conducting tests, or collecting data. In manufacturing, they may assist in product design, development, or production.

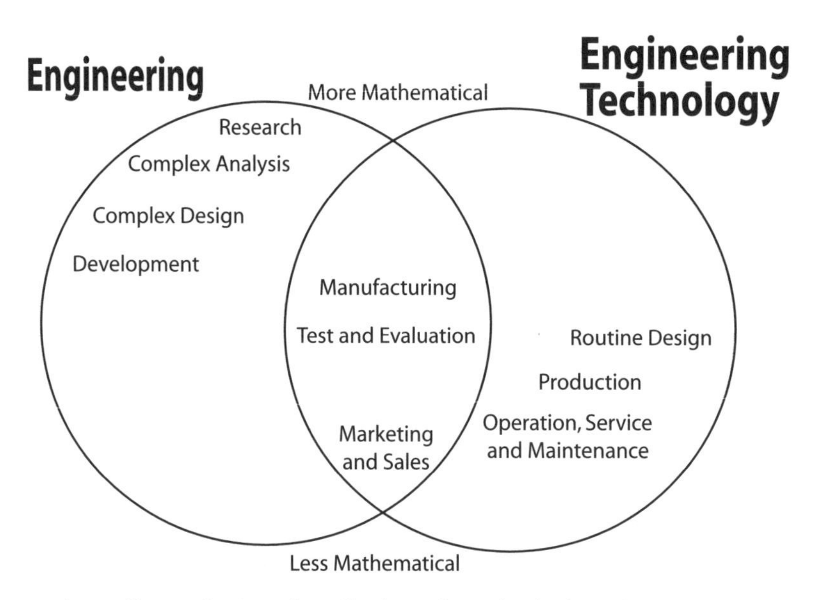

According to the American Society of Mechanical Engineers (ASME) program differences are delineated by the figure above showing clear distinctions and overlapping areas.

Engineering technicians who work in research and development build or set up equipment, prepare and conduct experiments, collect data, calculate or record results, and help engineers or scientists in other ways, such as making prototype versions of newly designed equipment. They also assist in design work, often using computer-aided design and drafting (CADD) equipment.

ET is a field that concerns itself with the application of technology. A technologist is an expert at applying technology to solve problems. An engineer may design a product to solve a problem, but the technologist will build, install and maintain it. There is some overlap in the two fields. Occasionally, with additional coursework, the technologist will also design a product, but most of the time, that is the work of an engineer. Conversely, some engineers are also involved in the build process, but most of the time that is the work of a technologist.

For example, every piece of gear has a designer, a builder, an installer, and a maintainer. Engineers are usually the designers; technologists typically build, install, and maintain the gear; and the technicians usually install and maintain the gear.

An advantage of engineering technology is that students can hit the ground running when it comes to getting a job after graduation. Employers are happy to hire technicians and technologists because of their training has given them the knowledge, hands-on skills, and ability to succeed in the workplace with minimal training. A technologist with a bachelor's degree can take the Fundamentals of Engineering exam - the first step toward becoming a professional engineer.

Other advantages of ET degrees include:

- Less time in college (this option saves money too).
- Not as much math and science.
- Great salary right out of school.
- Abundant job opportunities worldwide.

Disadvantages of ET degrees include:

- It's difficult to transition from a technician to an engineer because the science classes in an ET program are not calculus based.
- You may find that many years down the road, there is less flexibility in the career and less opportunity for advancement unless you seek advanced training.
- Salary increases may get smaller over time.

An advantage of engineering is that over the long term, you can climb higher on the totem pole. Engineers more often escalate to management positions and earn more over the life of their careers. If a career in research is interesting, an engineering degree can open the doors to further study. A terminal degree from an ABET accredited college in engineering is a doctoral degree, whereas a terminal degree in ET is a master's degree.

Advantages of an engineering degree include:

- More room for advancement.
- Easier to continue to graduate school.
- Great salary right out of school.
- Education is very broad – engineers can also become doctors, lawyers, writers, teachers, and business people.
- Abundant job opportunities worldwide.

Disadvantages of an engineering degree include: The work can be stressful – especially when the equipment has the potential to save lives. For example, in designing a new medical device, the project must be done to specification, on time, and on budget. If something goes wrong with the design in a few years or even 20 years down the road and it threatens the life of a patient, the engineer's job may hang in the balance.

Music engineering: is it right for you?

There are many approaches to combining engineering, technology and music. In general, a career in music engineering or technology requires that you be musically inclined as well as technical and creative. If you love music, like to work on computers, are fascinated by electronics and mechanics, or have a love for gadgets, combining music with engineering or technology may be the hot ticket for you. Not only can it lead to a successful career contributing to the newest releases on the charts, but it can also lead to success creating instruments or changing the way we listen to music. The good news is that this field is wide open with plenty of opportunities for a hard working ambitious person. You can school yourself at full tilt and get a formal engineering degree, you can get certificates or degrees in music engineering technology, or you can be self-taught. You can go to school for four or more years, two years, one year, one month, one week or one hour. Whatever your passion and whatever your taste, there is a flavor out there for you.

Engineering is not easy and the hours are long. In the music industry, the competition is fierce. Expect to start at the bottom and know that it often takes patience to deal with musicians and artists. Traditional engineering school teaches you to be a linear thinker. Making music is a creative process that is seldom linear. Making music or music technology equipment is a combination of art and science. You must be technically adept as well as creative, passionate about what you do, good with people, and willing to work hard.

When you love what you do, though, the hours won't feel as long. You'll look forward to going to work each day. When the little voice inside your head and every cell in your body says that this is it, you'll be happy and feel as though you were meant to be a music engineer.

There are four basic ways you can approach your career in music engineering.

1. You can become an engineer by attending a traditional engineering program and receive a bachelor's or master's degree in audio, music, computer, electrical or mechanical engineering (University of Hartford offers a mechanical engineering degree with an acoustics concentration.) Many students go on to graduate school - especially those that are heading into electrical engineering within a corporation or specializing in music engineering or

digital signal processing. Some students eventually earn a doctoral degree in electrical engineering or acoustics.

Audio engineering can be much more than music. It can be almost anything that we hear on film, tape, CD, DVD, computers, digital music players, handheld PDAs, radio, optical devices, telephones, cell phones, audio assistive devices, MIDI instruments, alarm systems, video games, thrill rides, etc. Some of the more challenging projects may be related to music, but there is also ample opportunity in a large variety of industries. Some programs will be offered through the music school and some programs will be offered through the engineering school.

To be an audio engineer usually means that you are interested in design. Education programs that offer bachelor's degrees to train engineers will be focused on design and building technical skills.

An audio technician will usually hold an associate's (two-year) degree or a certificate (one-year program) and will be interested in the implementation of the engineer's design. The technician will usually enjoy the practical implementation of building and/or maintaining the systems and designs.

Theresa Leonard, Chairman of the Audio Engineering Society Education Committee says: "The job possibilities in audio are growing in unexpected areas, such as game audio, car audio, and Internet audio, to name a few examples. A good program must look toward the future with respect to upgrading equipment and main control rooms for high resolution, new digital formats, and multichannel audio. Although it is important to know what is out there in terms of both high- and low-quality audio formats, there will always be an appetite for quality audio, which remains the driving force of our industry."

2. Get a certificate, associate's, or bachelor's degree in music engineering technology, audio technology, or computer music technology with a minor in electrical engineering, computer science, applied science, Web technology or multimedia.

   Are you a musician wanting a technology degree? Pursuing a degree in music engineering technology, audio technology, or computer music technology usually mean that you have an interest in one or more of the following:

- Studio engineering or design
- Studio installation, maintenance, or management
- Audio equipment design
- Electronic music
- Game or toy design
- Live sound for theatre and concerts
- MIDI production
- Ring tone design
- DVD authoring
- Software design
- Sound reinforcement
- Music recording
- Audio engineering or mastering
- Film, audio, and video production or post production
- Scoring for film and multimedia
- Synthesizer programming
- Broadcasting
- Audio sales

A minor in electrical engineering usually implies an interest in hardware design, whereas a minor in computer programming or computer science implies an interest in software design. When you complete your degree, you should be able to find a job as a recording engineer, multimedia author, sound reinforcement specialist, acoustical designer, studio designer, etc. The most common places to work include recording studios either for performers or soundtracks, audio equipment design or manufacturing companies, studio design companies, or for yourself as a freelance recording, mixing or mastering engineer.

3. Get a bachelor's, master's, or doctoral degree in music with a minor in electrical engineering or computer science.

   If music is your true love and being a musician rocks your world, there are dozens of schools that you can attend to follow this passion. Typically, people that pursue music and supplement it with physics, engineering, and math with find themselves either working in audio research developing better audio for portable devices, working in a recording studio, or working for an equipment manufacturer developing sounds for samplers or grooveboxes. In music programs, students learn the theoretical and combine it with hands-on practical solutions. This program develops a keen ear for music and can make you one of the most important and highly desired types of engineers. Programs range from studying in a conservatory environment to working on breadboards (electronic circuit boards) doing diagnostics.

4. Get a certificate, associate's, or bachelor's degree in recording engineering.

   Essentially, recording engineers store and retrieve music that comes in directly (MIDI instrument plugged directly into the console) and over microphones. They place the microphones, record the music onto tracks (digital and/or analog), and work the console. Recording engineers may record all kinds of music, including classical, jazz, rock, hip-hop, and rap. Some people decide to become recording engineers for the party. They imagine themselves listening to great music all day and chatting with rock stars. If this is you, listen up:— Your job as a recording engineer is to feel what the music is trying to say and help others feel it, too. It's not just about knowing what each knob does or understanding the equipment, it's also about capturing the sound—be it sound effects, speech, or music. Even if you don't like the music or what is being said, your job is to be true to it and work with the producer to help develop the expression of the artist. You have to get into it, open up, and remove your personal biases. With only about 50 major studios in the United States, this field is very competitive.

Recording engineering is a world in itself and many books are written about becoming and succeeding as an engineer in the recording business. This area is so broad and the information is so diverse that an entire chapter is dedicated to it. The approaches to careers as a recording engineer may be very different from the approach for a traditional engineering student seeking an audio engineering degree. For more information, see Chapter Two.

Actually, there are many more programs than those listed above. Degree and certificate programs come in all shapes and sizes. For a listing of all the colleges and degrees offered, see the school directory in the appendix.

Education options

The education you require is dependent on what you want to do. To be a music engineer, some folks say that getting experience is all you need. Others say that it's a combination of education and experience. There is no standard and no fail-safe road or path. Many companies require a four-year degree for an employee to be salaried (white collar). Depending on the company, they might not be very particular about the nature of the degree. Some companies have a rule that anyone with the job title of engineer has to have a four-year degree, but they don't care what the degree is. A few of the better engineers at top audio companies have degrees in English.

However, many schools are now offering hands-on training that is effectively narrowing the gap. Studios have to work harder to stay in business. Large studios are closing and being self-employed is fashionable. The most successful people in the audio industry have mentors, hands-on experience, and a certificate or degree to back up their portfolio.

Basically, college is designed to open doors. College gets you ready to go down the road toward employment. Entry-level engineers need some education in signal flow, signal processing, digital audio, console theory, and computer operating systems.

In music engineering, the average pay scale for a high school graduate is $29,000 per year. The average pay scale for a person with a bachelor's degree is $45,000. If you intend to one day support a family, chances are that you can't do it on $29,000. This author personally advocates education because it is always something you can fall back on. If you find yourself unemployed 20 years from now, you have some credentials that will keep you in the running against new graduates with skills on whatever new technology is available.

An audio enthusiast from Tennessee says, "I have been doing sound as long as I can remember and love every second of it. However, I'll be the first to say that you should stay in school. I didn't and wish I had. I currently work for an extremely large church, and do other gigs on the side. I stay pretty up to date on sound stuff, that's easy with three Yamaha Digital Mixing Systems in house as well as many other high-end digital consoles. I suggest you go to school and get a degree in something related to computers—be it computer science, networking, or something along those lines. Large studios are moving rapidly toward everything being on a network and being computer controlled. A good understanding of computers makes all these new features a lot easier to get up and running. Also, the piece of paper you get out of college is worth doubling or more your salary in the corporate world if you get into doing tech work for a major company, maintaining their conference rooms and auditoriums, etc. Most importantly, learn all you can, with technology progressing as fast as it is, the really cutting edge stuff that makes everyone's mouth drop open is all based around computers."

To get into broadcast, MIDI programming, gaming, and online music, or to get into digital music player manufacturing, electronic instrument design, and multimedia applications, you'd better get a degree. In such a competitive field, it's best to do everything in your power to set yourself apart. And, learn to play well with others.

Some audio programs may be based within another disciplines, such as electrical engineering, science, business, or music. Some programs will be interdisciplinary or a combination of all. For example, Belmont University in Nashville offers a degree in business administration with an emphasis in music production; the University of Miami's School of Music offers a Bachelor of Music, Bachelor of Science, and Master of Science in Music Engineering and their College of Engineering offers a Bachelor of Science in electrical engineering with an audio emphasis; Georgia Southern University offers a Bachelor of Science in Information Technology with a Second Discipline in Music Technology; the University of Massachusetts Lowell offers a bachelor's degree in sound recording technology for either computer science or electrical engineering majors; and the Berklee College of Music in Boston offers a Bachelor of Music in Music Production and Engineering.

Locating the right program for you will depend on your wants, strengths, and background. There are graduate programs, undergraduate programs, and numerous one- and two-year programs as well as one-day to one-week courses and online seminars in sound recording and music technology. You can find a curriculum that is ideal for your goals. How much money and time you invest is up to you and where you want to go.

When those of you in your early twenties were born, we had no CDs, DVDs or iPods. If we wanted to make a phone call away from home, we used a pay phone. If we wanted to see a movie, we went to a theatre. If we wanted to learn about something, we went to the library. We live in a designed world and engineers are on the cutting edge of this acceleration of society. By squeezing more information on integrated circuits or chips, the digital revolution is at full tilt.

There is no question that the digital revolution is also changing the face of music. Record labels used to be the only way for musicians to "make it." The labels would discover a diamond in the rough, produce their records, and then market and distribute the vinyl or CD. Now, any band with an entrepreneurial spirit can set up their own studio, record on MIDI and Digital Audio Workstations (DAW), burn to disk, distribute it on the Internet or sell their own disc at clubs and concerts. Some would say that engineers are becoming obsolete but anyone that has listened to good music will disagree. However, even if pushing a button to equalize a track becomes commonplace, there is still tremendous opportunity for recording engineers doing live concerts, setting up studios, designing instruments and gear, and in broadcast.

Choosing a school

Engineering is a demanding major. To be successful in engineering school, you will need certain tools. You must be self-disciplined and manage your time effectively. In college, the "real" learning often takes place outside the classroom, and less time is spent in the classroom. A general rule of thumb says that for every hour spent in the classroom, engineering students can expect to spend three hours outside the classroom, compared with two hours for non-technical majors. A good time-management system can also allow you to participate in extracurricular activities, which broaden your experience and are of interest to potential employers.

1. Location: In addition to distance from home, location refers to climate and the type of industry in the surrounding area. If there is industry specific to your degree, then opportunities for summer internships, co-op programs, and part-time work experience increase dramatically. These work experiences often lead to jobs after graduation.
2. Cost: Cost of attendance may be a critical factor in determining which school to select, although your decision should not be based on cost alone. Generally, public institutions are less expensive than private schools, but there are numerous ways to fund education at any institution. Scholarships, grants, loans, part-time work, co-op programs, and campus jobs also help reduce the cost of attendance. Check with the financial aid department of the schools you are interested in to see what you qualify for. Call the engineering or music department to find out about scholarships offered to incoming students through the college.
3. Faculty: A fine faculty makes it easier to get a good education. A faculty that includes under-represented minorities will broaden your experience and better prepare you to work with people from diverse backgrounds. Faculty members can bring numerous experiences and expertise to their lectures. Check to make sure that faculty has real-world experience as engineers, producers, studio designers, or musicians. Try to select a school that has at least one faculty member performing active research in your area of interest. That person can be a role model.

You can talk with and learn directly from someone whose interests you share.

4. School Size: School size matters for some students. Large schools offer a greater diversity of people and more things to do but often lack the professor-student interaction found at smaller schools. In a small school, you may get to know a larger percentage of classmates, but in a large school you can meet a much larger number of people. You can receive an excellent education at a large school or a small one; which you choose is purely a matter of preference.
5. Academics: Academics are probably the most important factor in choosing the school that's best for you. What programs do they have? A four-year or two-year engineering or engineering technology program should be accredited by the Accreditation Board for Engineering and Technology (ABET). ABET accreditation ensures that the school program follows national standards for faculty, curricula, students, administration, facilities, and institutional commitment. By choosing an ABET program, you can be sure that the faculty has met certain national standards and that the program is highly regarded by the profession. Schools of music should be accredited by the National Association of Schools of Music (NASM), and two-year technology schools should be accredited by the National Association of Industrial Technology (NAIT). Pick the atmosphere that best fits your personality and aspirations.
6. Facilities: Loads of studio time is usually very desirable in a program. The number of computers, workstations, MIDI labs, and microphones, as well as how much time each student gets on them are also extremely important. A school that is serious about teaching audio will have

invested in proper, up-to-date facilities. However, knowing how to work the programs and push the buttons won't make you an audio engineer. A student shouldn't pick a school based on the cool factor of knobs and buttons, but at the same time, it's easy to spot which schools have made the investment, and which haven't.

A good school will offer the fundamentals of audio as well as live recording time in different acoustical spaces, such as concert halls and/or studios. This recording time will help you develop the listening skills required to differentiate a variety of instruments while using a variety of different microphones. The school should also offer a critical-listening course. Ear-training packages can be purchased on the Internet, but nothing is better than feedback from your instructor and peers.

7. Talk to current students and graduates: Often the best resources about a program you are considering are the currents students and recent graduates. Schools should be willing to give you names and contact information, as this is part of their marketing. When you find the alumni, ask what they are doing now and how the school played into their achievements.

Find out if free tutoring is offered through the school and if the professors have posted office hours. Can you e-mail questions to professors? Will your questions be answered in a timely fashion? Prospective students can ask to sit in on classes. Another consideration is the campus library. Is it easy to find the information you are looking for? Does the school have a special engineering library or carry audio journals?

Other school selection criteria to consider include sports facilities, leisure activities, community events, cultural events, and campus activity resources.

Co-ops, internships and work-study programs

Cooperative education or a co-op experience is one where a student alternates work experience in industry with academics. For example, a student may do a parallel co-op where they work part time and go to school part time or complete a traditional co-op where they work for six months and go to school for six months. A good co-op program may be the perfect answer for the non-traditional student that has financial responsibilities.

Because a co-op program extends your time in college, the experience you obtain can be more meaningful. Additionally, a co-op experience can show employers you have experience and a solid desire to work in your chosen field.

Internships are another way to get your foot in the door. They generally consist of a summer job related to your major and quite often, you work for free. Co-ops, internships, and work-study programs should be viewed as a learning opportunity and give students the chance to find out if they really enjoy a certain type of work. They are a great way to get experience, and they may lead to industry contacts as well as jobs after graduation.

Michael Howes, Senior Acoustic Systems Engineer for the Logitech Audio Division says, "I don't believe in luck. I heard a quote once that goes something like: 'The harder people work, the luckier they seem to get!' This applies here. I worked my butt off in college. I earned a 3.87 GPA overall. During the summer I found jobs as a Multimedia technician, Bartender/Sound Engineer at a Jazz club, and even a 'roadie' with a local live sound company. It was hard work but it gave me experience on gear and I was in the industry. In my senior year I interned in the tech department at Sony Music Studios in Manhattan. It was a dream come true! I spent two days a week in Hartford and three days in the city. I aligned analog tape machines, helped repair gear, and even designed and built some gear using my electrical engineering skills. Before this internship, I had planned to focus on the recording industry for work. During my work at Sony, though, I realized how much fun it was to work on hardware. After graduation, I accepted a job offer from Logitech and have been with them ever since. Now I design microphones, headphones, and headsets. It's a blast!"

If you are interested in obtaining an internship, find a company you like and send your résumé as early in the school year as possible. Check with the job placement office at any school you are considering to find out the percentage rate, location, and contentment of placement.

An entry level position at Skywalker Sound

Following is an interview with Leslie Ann Jones, the director of music recording and scoring at Skywalker Sound, the recording and production facilities built by George Lucas. The interview appears in Keith Hatschek's book "How To Get A Job In the Music and Recording Industry," (available online from Berklee Press, www.berkleepress.com.)

K: Could you describe an entry-level position at Skywalker?

L: On the scoring stage, that gig is as a runner, which we have now although it's not a full-time position. The runner is just called in on an as-needed basis, because we only have the one music studio.

For the rest of the Skywalker facilities [home to the post production and mixing stages for hundreds of hit movies, as well as special effects division, Industrial Light and Magic], most people come in as central machine room operators (MRO) for the mix stage. Sometimes they might come in as transfer people, as well. But that requires quite a bit more experience and education.

A transfer op may have been somebody who worked at a smaller facility for a year or two, got their feet wet, and knows the difference between a single stripe and dual stripe mag, drop-frame and non-drop-frame time code, and so forth.

K: Could you identify three attributes or skills—it could be either—that you would look for in an entry-level person?

L: I think we tend to gravitate towards people that have the right amount of enthusiasm. We don't have a lot of people working here, and there isn't any formal time period that you're going to stay in each job. It just seems that those people that tend to excel at what they do, who grow and progress through the organization—start as a mix tech and progress to a mixer—are the ones that have the most self-motivation.

They can think for themselves, they are smart, and they invest the time to educate themselves. I really think not knowing too much and not knowing too little is key. I mean, even for a runner, the guy we have now studied for a number of years at Berklee College of Music in Boston.

I don't have to worry about him knowing the etiquette in the room or being unfamiliar with equipment. He has a really strong music background. Yet he doesn't know so much that he expects to walk in and be an assistant engineer right away. He's willing to make food runs and do whatever it takes to keep the session running, just so he can be here.

But there are only so many jobs, so you have to be flexible and be willing to fit in wherever you can. You need to stay attuned to the opportunities that might present themselves and be willing to jump in and take a chance.

That's what I've done in the last twenty-five years—let's see, I've had five jobs. This is my fifth job. And one of those jobs is counting the three years I spent as an independent engineer. I am pleased to say that, in each of my jobs, I have gone past what I thought I knew or tried something kind of different, with an element of risk.

Realizing that the next career move wasn't necessarily safe. That's the only way you can really grow. And that risk/growth relationship is a preview of what you're going to have to do when you finally sit in the chair as an engineer anyway. You are going to have to get past whatever knowledge you have to when the client says, "That's too orange."

You have to figure out what that means and how to make the track sound more "green." You should know enough about what you're doing and the tools that you have available to you to creatively get the job done.

K: What is the salary range for an entry-level position?

L: Well, interns get paid less than staff positions, although they do get paid. It is anywhere from $10–$18 per hour depending on what the person will be doing.

K: When a person is getting started in the business, they are there to primarily learn—not so much to earn. They need to try to get into a good learning situation, because the money comes later.

L: Yeah. Actually, that's why I really recommend that a person get a job in the biggest studio they can find and not take a job in a one-room place. Chances are, they're not going to really learn in a one-room studio.

K: Skywalker has an internship program. Could you talk about it briefly?

L: Well, it is handled through our human resources department. First a department like ours must decide each year whether or not to request an intern because the salary to pay the intern comes out of each department's budget. And then if anyone applied for an internship with the scoring stage as his or her preference, then we would probably get one.

But every company is different. Some do it like Capitol Records, where they would hire six interns from local music business programs throughout L.A., and they would spend a week in each department.

K: You mentioned you have a runner/intern now on the scoring stage. Can you estimate what percentage of new hires are current or former interns?

L: Around 20 percent.

K: Do you have any tips you can offer to somebody who is thinking about getting into the business? When you started, you walked in and approached Phil Kaye at ABC and said, "I'd like to be an engineer here." Things are quite a bit different now, obviously.

L: Yes, I think they are different. A lot of people that we consider tend to come recommended from other people in the business. We also have a relationship with certain schools. I might e-mail the head of the recording department asking if they have any outstanding students, which is exactly what I did the last time we were looking for somebody.

I contacted Los Medanos, San Francisco State, and Berklee College of Music and just asked if anyone had a couple of bright young kids. A referral like that is one way to get a start.

The other way is just to call around, and if somebody says they're not hiring, send them a résumé and follow it up with another call. Or, you can ask if you can come by, drop off your résumé, and see the studio. That way, the person who is hiring gets a chance to meet you, even though they might not be thinking about it at the time.

That approach may not work at some facilities that just do not have the time or the availability to accommodate drop-in visitors, but for many studios, it will work, so it's worth a try. You should ask, "May I stop by and drop off my résumé and meet you, and spend about five minutes speaking with you?"

Studio managers are generally very busy people, but at least you've had the opportunity to meet a person in the music community and hopefully make a favorable impression.

K: As far as resources, is there anything you think someone coming into the business should be looking at?

L: I think for somebody just starting out, "Mix" magazine might be a little too much. I guess "Recording" or "EQ" magazine might be a better place to start. We haven't yet talked about knowing computers, either. You certainly don't have to know Pro Tools editing, but you really should know the fundamentals of either a Mac or a PC. I think having some knowledge of hard disk editing is quite an advantage.

I would also suggest joining the Recording Academy as an associate or as an affiliate member, because you still have access to any of the workshops that are offered once you're on the mailing list. A lot of those events are free. So the networking and educational opportunities available in that organization are available whether you're a voting member or not.

A lot of schools have student AES (Audio Engineering Society) chapters; I know San Francisco State does.

As far as conventions, I would think now, NAMM would be a good place to go to learn a little bit about who the players are in the technology side of recording.

K: Long term, what's your sense of the career opportunity represented by becoming a recording engineer?

L: I think it can still be a [good] career opportunity, but I know that even well-respected veteran engineers are learning Pro Tools or some other hard disk editing system, because clients are kind of expecting that and they want that available to them.

Colleagues of mine have said, "Why should they pay someone else [to do hard disk editing,] when they can pay me?" So that's certainly job security. I'm still pretty bullish on that, but I think there are many opportunities out there with distribution changing with the Internet, uploading, and new technologies.

K: Do you have any parting thoughts?

L: Master the basics and the fundamentals. I think that's the big advantage of working in a big place and not in a small place—you're exposed to a lot more. In my nine years at Capitol, I was pushed to do so many things, not only the level of clientele that we had, but just kind of the things we were asked to do.

All the Frank Sinatra ednet ISDN sessions for the two Duets albums happened at Capitol. Then we shifted gears to record a film score with a large orchestra at the next session. You wouldn't really get that kind of experience in a one-room studio. It makes you much more valuable as an employee because eventually, you are going to have to look for another job. It always happens.

K: It's true. A person's depth of knowledge makes them much more valuable to their employer. Do you have any Yoda-like pearls of wisdom to share in closing?

L: Use your ears, Luke—use your ears.

Mentoring

Another excellent approach to become an audio engineer is to have or find a mentor. Mentoring is critical to success because it's a one-on-one learning experience that can be so much more than a technical learning experience. Mentors can help you learn approaches into this competitive industry, help you network, introduce you to key players, teach you how to listen, and evaluate solutions to problems as well as help you with technical problems. Mentoring is a vital part of being successful in any industry but especially in the competitive music business.

If you are considering a career in engineering, keep in mind that it will be a lifelong learning experience, and everything you do to prepare for it will help you reach your intended goal. The more you expose yourself to the world of music, technology, and engineering, the more opportunities you may have.

It is important to examine the shape of the career you want. Identify what kind of career you really want, and try to picture yourself in that role. If you agree to do a certain job or take on a specific project, make sure that you see it through to completion. If you don't know how to do something, learn how to find out! You can't possibly know everything about all subjects, but being smart often means knowing where to look for the answers.

Attitude is everything

Remember that in life, things can always go wrong. Murphy's law can also apply to any bad day, but the difference between a mediocre engineer an excellent engineer is their professional attitude and problem solving skills they take to work every day. A good professional attitude means being pleasant, personable, courteous, confident, resourceful, trustworthy, punctual, stable, efficient, and flexible.

According to the AES, good problem solving skills means: "Being curious, having the ability to work with delayed rewards, knowing when to work alone and when to work with others, knowing more than one path that will lead to completion of a task, knowing logical benchmarks and knowing how to evaluate progress and the ability to establish a broad and reliable network."

You can never begin preparing for this career too soon. Get involved in extracurricular activities that involve music, math, or science. Part-time, summer jobs or internships in recording studios are excellent experience. Call local studios and ask if they need help, and offer to work for free. Most studios are happy to help motivated students.

And remember that engineering in the audio field will always be a learning experience. Roy Pritts, Past Chairman of the Audio Engineering Society Education Committee, says, "An educated person is one who knows where to find the answers. No one can know everything, and it would be a useless effort to spend your life trying to know everything. The dynamics of change in our industry would find the rate of change faster than the rate of learning. The true challenge is to be in touch with the forces of change, evaluate those elements of which you must gain further knowledge, and place the rest just out of reach where you can turn quickly and acquire the necessary information when required. It is the difference between saying 'yes' when you know how, saying 'yes' then finding out how, or saying 'no' and being left behind."

Mark French, an assistant professor of mechanical engineering technology at Purdue University says, "Students must understand that the facts they learn in school will go stale pretty fast. They will need to stay current by continuous learning all through their careers. The part of their education that sticks is learning to think differently. Education is supposed to make you think in a fundamentally different way—they will enter an audio engineering program with enthusiasm, interest and maybe some experience. They should leave thinking like an audio engineer. Even if they forget the facts they learned there, the differences in thought will remain."

The Audio Engineering Society

Another successful strategy is to join the student section of the Audio Engineering Society (AES). The AES is the only professional society dedicated to the audio profession. The AES can be an excellent resource during your college experience and in your career search. They offer competitions against other colleges (check to see if the society has student chapters at the schools you are considering) as well as a journal, the ability to attend conferences at a discounted rate, tour facilities, use the section library, awards, guest speakers, and much more. The AES can also help you find internships, jobs after college, and summer work.

JL Rice

AES conference

The AES convention has mentoring sessions for students. During the convention, you can learn the benefits of mentor relationships, how to develop your own network of industry connections via mentoring, learn first hand from industry veterans and you can sign up for one-on-one mentoring sessions. Take advantage of it.

For details about the AES or student chapters and their activities, visit www.aes.org.

CHAPTER TWO

Recording Studios

Sound or recording engineer

Recording studio console

Recording engineering often seems like such a glamorous job. Most people envision themselves as recording engineers sitting behind the glass listening to great music all day, turning a few knobs and lounging with rock stars. They dream of travel, excitement, and fame. Although dreams are good to have, a dose of reality is also important.

In recording engineering, experience equals success. It may only take a few hours to learn what each button on the console does, but it's another thing to run the recording session. Just because you know what something is supposed to do doesn't mean that you can operate it properly.

The main responsibilities of a recording engineer are to record and sonically shape the music, translate the artist's vision into a recording, and ensure the best possible performance and sound in the studio. Recording engineers will do this by determining the exact placement of the microphones for the recording, placing the instruments and mixing headphones for the artists, and mostly, listening to and interpreting the goals of the artist and producer. Regardless of their music preferences, engineers must be able to technically translate the goals to the equipment and determine how it can best be used to produce the desired results. They must have good aesthetic sense, good musical sense, extreme patience, and a desire to always improve and expand their knowledge of recording equipment. In addition, they must be able to operate the recording equipment, know it inside and out, and also be able to dismantle and repair it. Good traits to get your foot in the door include being ambitious, creative, and relentless. Don't be afraid of approaching employers. Thinking about working in this industry won't get you the job—you have get out there, network, and keep your eyes open to opportunities.

Placing microphones is the first step in setting up a session. Sounds easy, right? Recording the sounds correctly is very important. For example, let's take the mike setup on the drums. Drum sets typically have seven or more cymbals and drums. Each drum needs to be miked accordingly. Many recording engineers place two microphones on the snare drum. The one above it get the "pop" sounds and one below it to get the hiss. They then place a mike at the bass drum and one at each tom (there can be up to seven of these). They then place two more for the cymbals, one for the high-hat, and then two more in the room to catch the ambient taps and boom of the drums. That's fifteen or more microphones, each streaming

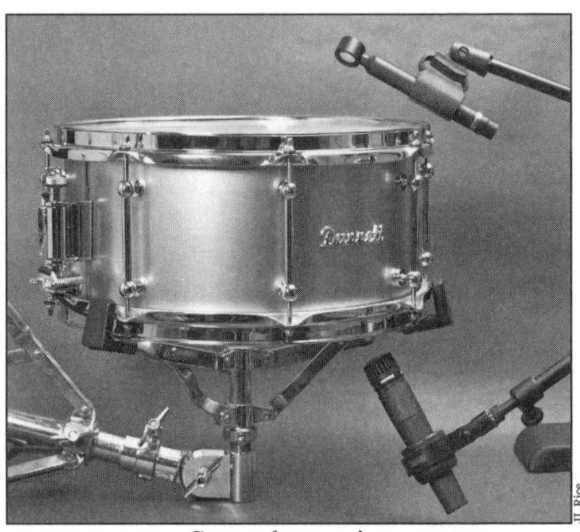

JL Rice

Snare drum mics

on a different channel, coming into the recording console simultaneously. Each can be individually adjusted for volume, effects, and frequency—and that only covers one instrument! Adjustments to drum sounds can be made to sound trashy, crispy, crunchy, or whatever texture and style the performers require depending on whether they are playing rock, rap, techno, classical, country, or jazz.

Once the engineer is convinced the microphones are placed properly and the sound is perfect, the recording begins. When the artists play, the engineer makes adjustments to the frequencies and effects to make some sounds stand out and other sounds more subtle. It's very detailed work.

Here's how it works: the engineer records the musician's songs one at a time. After the track is recorded, its quality is evaluated and discussed with the producer and/or artist. If it's not right, the engineer makes adjustments and rerecords the song. It can take several tries to get it right. After the engineer is confident they have a good recording, it is mixed down (balancing the volume and equalization of the instruments, vocals and effects) into two tracks that in turn, generate the master.

JL Rice

Guitar recording microphone setup

Engineers are the behind-the-scenes force. Not only is the engineering a vital part of the end product, it is also, in some ways, the most important. The job is a combination of high tech recording and creativity. It's a combination of right- and left-brain functions working together in harmony. Recording engineers must have an excellent knowledge of the equipment.

If a producer or artist wants a different sound, they must know exactly how to get it.

The hours are long, the work can be very repetitive but the creative side is what sets it apart from other jobs. For many people, that's also the best part of the job. The creative side isn't technical at all. It's more about the feeling of a certain tune. A good engineer is able to switch back and forth between the two cerebral hemispheres. Paying your dues as an assistant is part of the process for the awesome rewards of being the creative force in a technical project. You have to love this job to make the commitment. If you want a more structured life, where you start work at a certain time, end it at the same time everyday, and make it home for dinner each night, this might not be the right career for you.

Tasks for recording engineers include:

- Recording speech, music, and other sounds
- Equalizing the recorded speech, music, or other sounds
- Synchronizing the recorded speech, music, or sounds to the visuals of a music video, DVD, television show, or movie
- Mixing and editing the recording
- Mixing and editing during live concerts
- Regulating volume and sound quality during performances
- Regulating the volume and sound quality during music videos, DVD, television shows, broadcast productions or movies
- Keeping a record of recordings
- Maintaining the recording equipment
- Installing, adjusting, and testing recording equipment to prepare for a session

The cities with the largest number of recording studios are New York, Paris, Nashville, and Los Angeles. However, many cities across the country have smaller studios that may be a good place to gain some experience. The right choice is a matter of preference.

Assistant engineer

The assistant engineer is the bottom of the totem pole. It's all about paying your dues to reap the considerable rewards later. It's a hard job with unpredictable hours, stressful time crunch situations, and low wages. Caffeine and take out will become your friends as you work for long stretches of time to get your foot in the door.

The best preparation for a career as a recoding engineer is experience. Assistants looking to make it to the next level should take every opportunity to assist and observe the recording process. On the job, in a studio, at a concert, the choice is yours. Pay attention to detail, exercise patience in all things, and work on your communication skills. You can't begin preparing too soon for a spot in this competitive industry. Experience is just as important as formal education—so start early. High school students can find jobs at recording studios, record labels, music promotion firms, and music and record stores.

On the upside, for creative types, this is a dream come true. It's your chance to get away from the nine-to-five corporate world. In the process, your office will become a movie set, stadium, theatre, concert hall, or recording studio. You'll possibly be working with music legends or putting together CDs for new bands, or making a soundtrack for a new movie or video game.

Mastering engineer

Mastering is the creative bridge between the final recording and the manufacture of the physical CD, DVD, MP3, or any other format that is released for sale and radio play. Mastering engineers equalize the final mix, ensure consistency on different playback systems, put the songs in the desired order, edit or reconstruct the arrangements, and transfer the recording from the master tape or disc to a Master CD or the manufacturer's specified format for duplication. They decide the volume of each song, time between the songs and the ultimate sound. They work closely with producers and artists to make the music enjoyable for listeners. Once a mastering engineer gets a name in this industry, the compensation goes from standard to excellent.

Post production engineer

The post production engineer goes to work after the production is recorded. They are the final step after the mastering engineer completes their job and are responsible for what you hear when you watch TV, films or videos. The finished sound required for DVD is different from the sound required for movie theaters and is different from the sound required for MP3 players. This engineer makes sure you get the best listening experience for your format. Engineers in this position are highly technical and creative. Post production engineering has tremendous growth potential and very good engineers often get to see their names in movie credits.

Studio designer or acoustic consultant

Any home studio designer can tell you that music recording and production doesn't have to happen in an expensive state-of-the-art recording studio anymore. They'll also tell you that there is much more to it than meets the eye.

The goal when creating a music studio is to create a near perfect environment that sounds great and is also dead quiet.

Sounds from the street, outside hallway, barking dogs, etc. can ruin hours of hard work and drive a studio out of business. Special, innovative materials and installation techniques have to be used because the requirements for a studio far exceed standard construction techniques. Imagine that you have a recording studio in your garage. Bands typically play from 100-110 decibels (dB). Inadequate sound proofing not only breaks most city sound ordinances, which are in the neighborhood of 65 dB, but many frequencies will be lost and outside sound will ruin the recording.

JL Rice

Home Studio Recording Equipment

To give you an example of the challenge of studio design, according to Quiet Solution, the maker of QuietRock, a quiet post-production room will have a maximum of 25 dB and a quiet recording studio will have a maximum of 20 dB. A dB, is a unit measure reflecting "how loud" a sound is. Decibels can be measured using a dB meter. For comparison, whispering is about 20 dB, a vacuum cleaner is about 80 dB and a jackhammer is about 100 dB when you are standing seven feet away from it.

The Sound Transmission Class (STC) rating of a wall is a way of knowing the sound transmission loss through a wall. The higher the STC, the greater the sound attenuation through the wall or barrier and better the sound. In most houses, the minimum building code requires an STC of 50 and STC of 60 to 72 is considered excellent.

Not only does this type of designer need to understand acoustics but they also need to have a good feel for aesthetics and know about innovative building materials for sound absorption and deflection. Outside of computers and electronic equipment, primary considerations for studio design and construction include sound isolation, reverberation design, vibration control, and heating, ventilating, and air conditioning (HVAC) noise control.

TV engineer

The following was reprinted from High Tech Hot Shots: Careers in Sports Engineering, www.engineeringedu.com.

My name is Cindy Hutter and I work for the Fox Sports Network at the Houston Operations Center—a 100,000-square-foot, white metal building in an industrial neighborhood just outside the famous Galleria area in Houston. You wouldn't know much happened in here from the outside. Inside is one of the largest technical television program playback centers in the country.

What we do at the Houston Tech Ops Center:

- Take in tapes from program suppliers for the purpose of broadcasting them on Fox's regional sports networks.
- Receive live professional and college sporting events via either fiber optic line or satellite to broadcast them on the regional sport networks.
- Add commercials and graphics to the live and taped shows.
- Digitally compress 14 program streams into 5 satellite transponders to permit their broadcast to some 8,000 receivers at cable company "head-ends" all across the country.

The Houston Tech Ops Center has 23 fully digital master control rooms, all run by an automation system. At any given hour, one operator controls three networks at a time. Most recorded programs are played to air off one of ten video servers, with a total storage capacity of 15TB.

Outside the facility are the technicians who work in the "live trucks" at the ball fields to produce and transmit the games back to Houston. It is these technicians who have to know the most about sports to be able to work in the business. The rest of our broadcast engineering work is rather universal—it's the same across sports, news, commercial production and entertainment.

My primary responsibility is to oversee the entire process of the Technical Center—300 employees over eight departments ranging from graphics to the video library. It is my job to find ways to make the broadcast process more streamlined both from a technology and manpower perspective.

I have degrees in Electronics and Journalism with a focus on Radio and Television operations. Many large-market television stations like to hire Directors of Engineering with a BSEE (focus on Electronics), but many people in television engineering do not have engineering degrees (or college degrees at all!). While an engineering degree and the experience it brings are helpful, often "OJT" is just as good, and is often preferred.

I definitely had a lucky break. I graduated from college in the mid-1970's, when female technicians and engineers were difficult to find, and TV stations were hiring all of them that they could find. In 1979, I got a "summer vacation temporary engineer" position at ABC News (and spent the summer in the electronic maintenance shop fixing TV Monitors and soldering multi-pin connectors on camera cables). I was able to stay at ABC for a number of years, working in all the technical areas gaining a lot of experience, which included exciting engineering assignments in locations like Brazil and the Himalayas and all over the world. This background made it possible for me to later attain senior engineering management positions (including several stints as Director or VP of Engineering) for several television stations around the country. From that background, I was able to attain my present position at the network.

As you can see, it is extremely important to find an entry level job, such as my paid summer temp job (a lot of TV stations and other TV companies hire extra technical help in the summer time so that the regular engineers can take vacations) or an internship at a television facility in order to "get your foot in the door". Real-world experience makes all the physics and math come alive, and you gain an understanding for how video, audio and data move around a television facility and the principles behind video flow—something that it's very difficult to teach in a classroom. Another possibility is to work at a local cable access channel—many won't pay much, but it's invaluable experience.

Adam Gehrke of KPLU radio and Q13 Fox News in western Washington

I was lucky enough to know what I wanted as a career soon after I started working at the student radio station in high school when I was 15. I've worked in radio, concert mixdown, and since graduating college, television broadcast engineering.

Online music engineer (AOL music, iTunes, Napster, Webcasts, etc.)

Becoming an online music engineer or Web audio designer is a career that no one would have thought about just 20 years ago. The explosion of music on the Internet has increased the opportunities available to audio, electrical, computer, and software engineers in this arena.

Engineers in this field may develop large-scale Web applications, relational databases, program in Java, PHP, coldfusion or other Web-based applications, and they prepare audio files for Web channels. To be excellent in this field requires an almost intuitive understanding of how people listen to music, and an uncanny ability to solve problems (fix bugs). You'll also need to anticipate how to improve the customers' listening experience and possess solid written and verbal communication skills to relay your suggestions. You may be responsible for training, testing functions, content management tools, or Web publishing applications, documenting procedures and/or developing new technologies. Exceptional computer and Web music application skills are also highly valued.

Corporate studios (CocaCola, FedEx, IBM, GM, Boeing, etc.)

If you are focused solely on making CDs for rock stars, you may be missing out on some fantastic and lesser known opportunities to work in audio. For example, many large corporations have their own in-house studios. Hotels and companies maintain A/V systems and produce internal presentations. Many large companies are racing to produce videos or Web productions to entice customers. A student with knowledge of audio and HTML may be an excellent fit for this industry.

Amusement park audio (thrill rides, animitronics)

The following was reprinted from The Fantastical Engineer: Careers in Theme Park Engineering. www.engineeringedu.com.

There is no more effective tool for shaping the mood in a space than sound. Consider the feelings you experienced when you last heard the following movie theme songs: Mission Impossible; title song for Star Wars, "Imperial Death March"; James Bond 007; Spiderman; The theme song for the television series "Friends"; and Rocky, "Eye of the Tiger."

Roller coaster sound effects are used in a variety of movies.

In many respects, these popular movie and television theme songs are just as recognizable as the shows themselves, if not more so. Just as no television show or movie would go without background music, the power of sound should never be neglected as a mood-enhancing tool for themed attractions either. Sound is all-important, whether it is musical theme songs, special effects, or story enhancing dialogue. Too, sound should always be over-utilized rather than under-utilized. Sometimes, silence is desired, and in such cases it should be used to punctuate the drama of a situation, but never because "we just couldn't think of something to do there." Just as television and movies continuously use background sounds to add mood and interest, so should it be with themed attractions and architectural showplaces.

The next time you visit Disney's Animal Kingdom, be sure to pay attention to the relaxing mood music continuously played in several key areas of the park, especially near the front ticket gate. In my opinion, the music helped contribute to a relaxing themed experience even on the most crowded days. Indeed, I visited Animal Kingdom on a day when they set an attendance record, yet I noticed that I did not feel stressed out like I would normally under those circumstances. I attributed my relaxed mood that day in part to the presence of the background music.

CHAPTER THREE

Live Sound Engineers

Concerts

Live sound engineers, also know as Front of House (FOH) engineers, are multimedia experts that are responsible for the sound and dispatch in all live shows—concerts, clubs, churches, sports events, presentations, festivals, plays, musicals, etc. Typically FOH refers to engineers who work the consoles in concerts. Live sound engineers may also be known as remote engineers if they work on location at live events to broadcast an event such as the Grammys or the Academy Awards. They may be known as Foley engineers if they work primarily on Broadway designing the sounds in a performance or they may be known as sound reinforcement engineers.

J.L. Rice

Outdoor concert

This type of engineer is required to have a wide range of skills. Not only do they need to understand the math and physics of sound design, they must also understand music, be good communicators, and be good trouble shooters to keep the gig running smoothly. Most of all, they must like to travel. Although it is understandable that this engineer may only be hired for one night, they may also be hired to tour for a few years. Phil Clark, an FOH mixer says, “Only a select few (from what I can tell) make the ‘Big Money’ on tours. But like the artists themselves, those gigs can disappear overnight in some cases. I agree with the idea of getting a ‘real’ degree in something that you can have a real future in, and that way if the sound work dries up you have something real to fall back on.”

In a concert setting, you have the best seat in the house behind the mixing console. You are responsible for providing the best sound experience possible. The trick to mixing in a concert venue is to make sure that no matter where you are in the crowd, the listening experience is the same. The people in the balcony paid a good price for their ticket and deserve the listening pleasure of a great show.

Not only is this type of job travel-heavy and very creative, but you have to be a top-notch problem solver too. FOH engineer Mark Amundson, says, “There are hundreds to thousands of sound persons in every state, all dreaming the touring fantasy. But only a lucky few will get picked for this duty. And it is very much a young person’s occupation, due to the physical demands and amount of travel involved.” You may be in a new place every night, have different crew members installing or tearing down equipment. You may have gear compatibility issues, and dealing with different artists isn’t always easy. Instruments can fail, gear can be faulty, and sometimes you may have to do pitch correction if the lead singer is having a bad day. You never know if you will be required to do some last minute cable soldering or if the band will try to change mic positions or amp setups in the last minutes before the show. Whatever the case, you must be ready to adjust your setup accordingly.

Most companies that look for an FOH engineer want someone that can play well with others. It’s your job to be customer service oriented (both to the gig manager and the audience), keep the peace, and make sure the show goes smoothly. You will be expected to be quick minded and fix any sound problems on the fly.

Many FOH engineers say that they love digital consoles because they can run many feeds simultaneously. The live feed might be for the house (live audience), then there could be a radio feed to broadcast it live, an Internet feed for a streaming concert, and a video feed so everyone can see the stage and watch the concert unfold. In addition, the engineer must be thinking about how the sound will be used later. Will live CDs be made from the show? Will any part of the show be sold as footage for VH1 or MTV? How the recording will be used is just as important as getting the right set up for the real-time live show experience.

Digital consoles are also great for maintaining control. John Shivers, FOH engineer for Tarzan says, "Every orchestra reverb is done within a PM1D Digital Mixing Console. We have a drum reverb, a guitar reverb, a guitar flanger for the acoustic guitar, etc. There are eight different effects available there, so you can dedicate them specifically to sections of the orchestra and maintain a lot of control. If you want a gated reverb on the snare, you can have it while having other reverbs on other sections. That's another beautiful thing about the digital console, a vast range of possibilities on a cue-by-cue basis. If you're changing EQs, changing noise gate configurations, compression thresholds, or just about any parameter, it can be done at the push of a button. That really makes the digital boards very, very powerful and very enticing to me."

Churches

Church music is big business. Many houses of worship are doling out the money to install State of the Art sound systems. These churches are a dream come true for the FOH engineer/wizard of gospel. One church claimed that it had an 11 piece band: drums, percussion, two keyboard players, guitars and brass as well as some praise singers, and a 200-voice choir. With 40 microphones for the vocals, this church is running almost 100 channels for one service. In addition, the church is likely to have a large mix of monitors (speakers and amps), wireless microphones, digital consoles, delay systems, a line array system, special effects, and triggered lights. And the church is likely filled with acoustic treatment.

These hard rocking houses of worship are taking a modern approach to attracting their congregation. We live in a world that is highly technological and stimulated by sight and sound and the church does not want to be left behind. Many pastors might even be speaking to a radio or television audience as well as to the congregation. And why not? Church composers, musicians, and singers are highly talented and have the ability to mesmerize an audience with their words and message of hope. Using music to attract and retain an audience is powerful and alluring.

On Broadway – Musicals

Mixing sound for Broadway productions comes with a variety of unique challenges for the FOH engineer. In Broadway shows, the FOH engineer is primarily concerned with creating a smooth and natural sonic environment that engulfs the audience and draws them into the production more than they normally would be. These engineers must have a good understanding of how to record voices, set up a sound system, and be excellent communicators. You must know how to deal with the different acoustics in each theater, the different needs of each show, and the different people that run them. In this line of work you may also work closely with a Foley engineer, editor, mixer, or walker to create all the little sounds for a play or musical such as a door closing, a car starting, or a coffee pot brewing.

Think back to a time that you went to the movies and the surround sound was either not working or disabled. You probably couldn't feel the movie as much...you couldn't hear the jets flying around your head. The experience of seeing the movie wasn't as riveting as when the sounds were surrounding you in the production. With that in mind, the FOH engineer must work tirelessly to create just the right experience in every different Broadway play. For example, in the production "Tarzan," the engineer had to be concerned with making the audience feel as though they were in the middle of the jungle. The jungle had to incorporate the background sounds of owls or crickets during the night and birds or frogs during the day. This could not have been believable without the sound surrounding you from right, left, front, back, and overhead (jungle canopies). In addition, the silverback gorilla had to roar occasionally and needed vocal processing (using a pitch shifting program) to embellish the actor's voice.

Recording the sound of brewing coffee

Some engineers say that the hardest part of the job is the long hours during a production. Shows can run six days a week for six months. You may be on the set from 8 a.m. to midnight, mixing rehearsals or the show itself. You may be responsible for the MIDI show controls (light board and triggers), computers and projectors. In addition, as the show moves around the country, it may change scale. The show in San Francisco may be very large and different than the smaller show in Seattle. People in San Francisco also might react differently to the sounds in a production than people in New York.

Film and broadcast engineer

In film and broadcast, audio engineers either record voices/ dialogue, music, or sound effects. There are positions in production and post production.

Recording during production means capturing the sounds in real time. Production work includes placing microphones and either recording voices in a studio doing an animation movie or you may do this type of work on location. On location, you would be required to capture the ambient sounds of a natural setting (birds chirping, frogs croaking, or a snake slithering) as well as the conversation of the actors or actresses.

Working in post-production means that you work on the film after it has been captured. For example, in a car commercial, you may find that you can enhance the understanding of the viewer by including the sounds of a skidding car to show how well it stops or the sound of peeling rubber to show how well it goes. You may find that a scary movie is scarier if you add the sound of the wind or clanging of chains. You may work directly with a sound effects editor or you may be the sound effects editor.

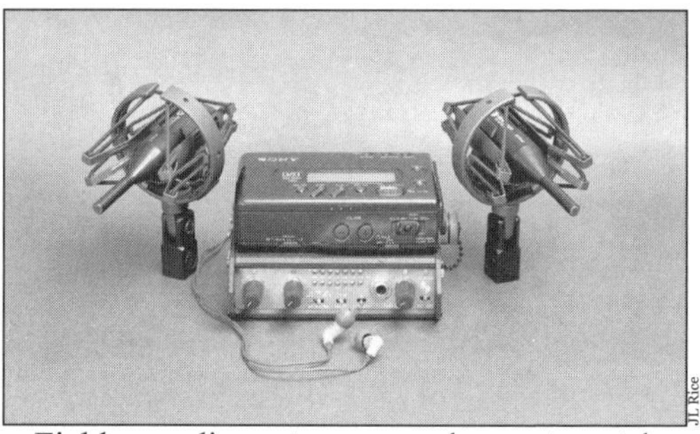

Field recording setup — can be use to catch ambient sounds.

To do music work for film or television, you would work closely with a music editor and composer. You may be involved in placing the music from a jukebox or car radio that is playing on the screen into the film or show. If you like to work in a fast paced environment, TV may be the way to go because the show will air every week. TV and film jobs are usually union, so you will be able to find the job description and pay scales online.

Concert lighting technicians

The concert lighting technician is a master at producing lighting effects that are emotional, complex, and spectacular. Today, in many live sound applications such as live theatre, concerts, corporate events, laser light shows, festivals or clubs, the lights are synchronized with the sound and emphasize the mood of the music. Lighting effects that have tremendous visual impact can be computerized or mechanical. The lights or pyrotechnics for a concert can make or break a show. Your job will include working closely with the FOH engineers, event managers, and producers. You will also travel extensively.

The demand for lighting technicians has increased due to more corporations holding events and the influx of multimedia permeating every entertainment venue in our society. A career as a lighting technician could easily graduate to a career in lighting design or as a lighting director. If you are good with special effects and/or enjoy handling fire, a career in pyrotechnics could also be for you.

The work can include rigging the basic stage lights or spotlights as well as using the computer or control boards to create special laser or lighting effects. The technician usually works closely with the director to make sure the lighting effects are provided on cue and the show produces the "feel" the director was striving for. In addition, the lighting technician arranges for the transportation of the lights and ensures that all lighting is in safe working order.

In general, lighting technicians, under the direction of a lighting director, designer, or stage manager may:

- Make sure all the lights are in good working order
- Run the cables and electricity to the lights
- Set up the lighting equipment
- Program the lighting console or control board (synchronize with the music)
- Operate the control board during a show
- Set up generators to power the sound equipment and/or cameras
- Dismantle and pack up equipment after the show

Lighting technicians should have a good working knowledge of electrical systems and electronics. In addition, they should be:

- Patient
- Creative
- Safety minded
- Good communicators
- Good problem solvers
- Able to work quickly and efficiently
- Able to work at heights and physically fit (have stamina)
- Able to work in confined spaces such as lighting booths
- Able to work in all weather conditions for outdoor concerts

CHAPTER FOUR

Instrument Design and Manufacturing

Instrument design and research

The market for instruments varies with the economy and school budgets, and although the competition is fierce, there is still opportunity in this field. But be forewarned, without knowledge of computers, MIDI and software, this field is extremely tight.

To give you some history, in the early 1900s there were almost 500 piano manufacturers in the United States. Today there are only about ten. In the 1980s, the United States imported almost five times more instruments than it exported. The future in instrument design for an engineer is no longer that of an artist that handcrafts instruments, the future lies in developing new and improved MIDI systems for all

instruments as well as improving the current applications or capabilities of instruments, music equipment, and accessories.

For example, engineers in this field might work on developing more authentic drum sounds for vDrum kits or improved pickups for electric guitars. They may be working in the development of a new and improved electric violin or MIDI saxophone. Companies that make amplifiers, speakers, microphones, sequencers, mixers, etc. also hire engineers. There is plenty of opportunity in this field for the computer savvy engineer with a love for converting the analog sounds of any instrument to the digital world.

Most engineers that work in music research labs have music training or a background in music. Not every job in this industry requires an engineering degree (such as tool and die makers in the production line), but if you want to do any research or design, it's best to get an education.

Designing musical instruments is creative and artistic. The industry is small compared to other manufacturing industries but none the less, engineers are still required to invent new musical instruments as well as modify and improve old designs. Because so many instruments are moving to include MIDI capability, it is an especially good time for engineers with electronic and software programming knowledge to jump into the mix.

Drums

Drums are known be the second oldest instrument in the world; right behind the human voice. As you probably know from tapping on different objects throughout your life, there are many factors that can change the sound of what you hit. With traditional drums, it doesn't matter whether you are looking at a bass drum, a snare drum, or tom-toms, the elements that affect the sound are the size, shape, thickness and type of shell material. The type and weight of the stick, type of stick tip material, and hands, as well as the drum head treatment (damping material) also greatly affect the sound. Other considerations that affect sound are the heads and the gear hanging from the drum. (For example, tom-tom holders attached to the drums can reduce the vibration and deaden the sound.)

In general, the larger the diameter of the drum, the lower the pitch. This is especially evident with tom-toms that come in sizes from six to 20 inches in diameter. The pitch of each drum gets lower as the drum diameter gets bigger. In addition to the diameter, the sensitivity can be adjusted by changing the depth of the drum. A skinny drum will have acute sensitivity, so the musician will be able to play quietly with full tone vs. a fat or deeper drum that will have more range on low sounds, but they will require hits with more force to get the full tone.

JL Rice

Variety of tom-toms

Drums are usually made from several layers of wood. The most common wood is maple, but birch and mahogany are also popular. Some companies are also using acrylic, fiberglass, or composite materials, such as Acousticon (resin-impregnated wood fiber) but musicians complain that although it projects better, the sound lacks warmth. Thinner shells may be only four layers and resonate a "woody" tone. This is good for recording. A medium-thick shell is usually six layers, and a thick shell is eight layers. As the shell gets thicker, there is less vibration, which allows the sound to project further. You'll find that really loud bands use thick shells.

Snare drums are the exception to this rule in that their shells can be made of brass, copper, steel, bronze, carbon fiber, stone, plexiglass, stainless steel, tin, ceramic, aluminum, or wood. Each type of material changes the sound. For example, brass will produce a brighter sound (higher pitch) than wood, which may be good for a marching band, whereas the dark and warm sound of the maple shell may be required by a jazz band.

JL Rice

Variety of snare drums

Electronic drum sets are another option for the drumming engineer and have taken the world by storm. Suddenly, you can bang away on a drum set in the middle of the night and not wake up your parents or the neighbors. Basically, engineers took samples of hundreds of drum sounds on every different type of drum, put them into a computer database and then attached that computer to the drums. Viola!

Electronic drums are made out of rubber and act as "triggers". That means that when you hit the pad, the sound of the sampled drum is generated and you hear the sound. This is very convenient because the computer can hold samples of any type of drum set. Instead of having a drum for jazz music, a drum for rock, and a drum for hip hop, you can simply select the drum you want on the electronic kit. This is ideal for the traveling musician that doesn't like to lift heavy drums. The electronic drum kit industry is a great fit for a drumming computer, software, materials, mechanical, or electrical engineer.

Besides traditional drums, engineers may also work in cymbal design. They may develop new bass drum or high hat pedals, triggers (which allow electronics to be hooked up to a traditional drum set), sticks made of new materials, and any other percussion instrument or gadget a band may require.

Guitars

The electric guitar is one of the most amazing blends of creativity, electronics, craftsmanship, and music. When you listen to rock and roll, the guitar is one of the most popular instruments. The guitar sound (or the guitarist) can make or break a band. Invented in the 1930s, the electric guitar has made its way into the hearts and souls of music lovers everywhere.

Guitar engineers come from every imaginable background. Mechanical engineering (ME) with an electrical engineering (EE) minor (or vice versa) is a very good choice. A physics degree with an EE minor is also a good choice. Whatever your major, make sure you have solid electronics training. EE is very common but don't be misled because although the name is deceiving, electric guitars actually don't require electricity. You don't have to plug them in and they don't run on batteries. The vibration of playing is what causes the guitar to produce electricity that is sent to the amplifier. Electric guitars can have solid or hollow bodies. They are similar in shape to the acoustic guitar in that they have a neck, frets, and tuning pegs.

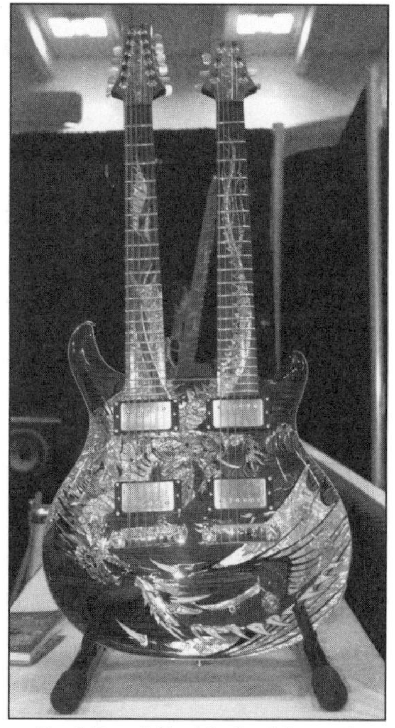

Paul Reed Smith double neck guitar

However, a major difference in the electric guitar is the magnetic pickups that are under the strings. The bar magnet/coil picks up the vibration that is produced when you pluck the strings and sends it as an electronic signal through a circuit on the guitar and then out to the amplifier and speaker. Guitars can have two or three pickups that can be adjusted to produce different sounds. Engineers in this industry need to have a good understanding of circuits and magnetism. Although a four-year engineering degree is not required, an associate's degree in electronic engineering technology or the equivalent coursework is preferred.

If the concept of vibrations producing electricity baffles you, what is even more amazing is that our ears also produce electricity. Hearing is achieved when vibrations in the air move our ear drum, which in turn, moves the three little ear bones that are attached to it. This change in air pressure is picked up by our ears as sound. There is an electric charge in every cell of our body and so an electric signal is sent to the brain, via the auditory nerve, which tells us what the sound is.

Lloyd Loar, an engineer at Gibson Guitar Company in 1924, was the first person to develop an electronic pickup although it was on a viola. In 1928, Stromberg-Voisinet developed a pickup for a guitar. The problem with these early pickup devices was that the signal was too weak. It wasn't until 1932, with the introduction of the "Frying Pan," by George Beauchamp and Adolph Rickenbacker, that electric guitars became truly successful and captured the hearts of millions of music fans. In 1939, Les Paul created a guitar called "The Log" because he mounted a pickup on a four-by-four piece of pine. In 1950, solid body guitars became the rage because they didn't produce the feedback common in hollow body guitars. Leo Fender and Gibson both produced solid body guitars in the 1950s and 1960s, such as the famous Stratacaster, that were mass-produced.

Guitar tech (on tour)

Being a guitar tech is like always having a backstage pass. These technicians are onstage to make sure that the guitars are in good working order. Have you ever watched a concert and seen the guitarist trade guitars with a person on the stage? That was the Guitar Tech. The tech works to make sure that live performances go smoothly. They change broken strings, handle the packing and transportation of the guitars, as well as clean, polish, and tune the guitars. They are the guitarist's best friend when problems arise. If you like to travel, this may be the perfect gig for you.

The best training for guitar techs is to work for a guitar manufacturer. This gives you the ability to understand the dynamics of many different guitar models. For example, if a guitar is made of wood, heat and humidity can change the sound of it because it expands and contracts. A wood guitar played in Miami will sound different if played in Arizona. If the band wants a certain sound, the guitar tech has to know what guitar is the best choice for each location.

Speakers and amps

Amplifier and speaker design is another area in the music business that employs many engineers. Amplifiers are used in televisions, stereos, MP3 players, computers, automobiles, alarm clocks, and anything else you can think of that produces sound. Amplifiers increase the signal strength (sound) of your home stereo, electric guitar, electric drum set, keyboard, and any other MIDI instrument.

The job of the amplifier is to amplify or increase the sound waves or audio signal to a point where it can move the cone in a speaker back and forth. The amplifier increases the current so that the sound can be heard. That's why the amplifier for a rock band is so much bigger and stronger than the amplifier for your MP3

JL Rice

Near field monitor

player or clock radio. In addition, high fidelity amplifiers require precise engineering. There is much more to amplifier design than we will cover here. If you are interested in speaker and amp design, visit Howstuffworks.com for an in-depth description of transistors, distortion, and basic electronics along with excellent pictures.

In a nutshell, speakers take an amplified digital signal from a CD, DVD, TV, MP3 player, radio, MIDI instrument, etc. and produce sound waves that vibrate our ear drums (which our brain identifies as sound.) Speaker design is also a booming industry as we, as a society, crave better sound or simply blow out our speakers. We can have the best amplifier, but if our speakers are of poor quality, the sound will be of poor quality, too.

Microphone engineer

Microphones are like paint brushes. Each microphone has a specific application, and to get the right sound requires using the right microphone. You wouldn't use a skinny brush to paint a large mural, but which brush you use is a matter of individual preference.

Microphone engineers must have a good understanding of transducer design, electronics, and audiology. Microphones work by converting sound waves (vibrations) to electrical current. They have a very thin and extremely sensitive diaphragm that is vibrated by sound waves. This vibration or movement produces a small current that must be amplified to hear it through your speakers.

So, from the section above, you learned that vibrations in the air produce sound. In other words, when you speak into a microphone it produces vibrations that generate an electric signal that is sent to an amplifier. The amplifier increases the signal strength and sends it to the speakers. The electric signal, in turn, vibrates the cone in the speakers, which produce the sound.

Working for a microphone design company usually means working with many other people. Microphones can be omnidirectional, bidirectional, or unidirectional. They can be worn on the head or lapel, be mounted on a stand or handheld. They can be used by rock bands, football stadiums, orchestras, choirs, radio broadcasts, conferences, and any other entity that needs to be recorded or heard in the back of the room. With so many uses for microphones, engineers must have a good understanding of what is needed in each application and environment.

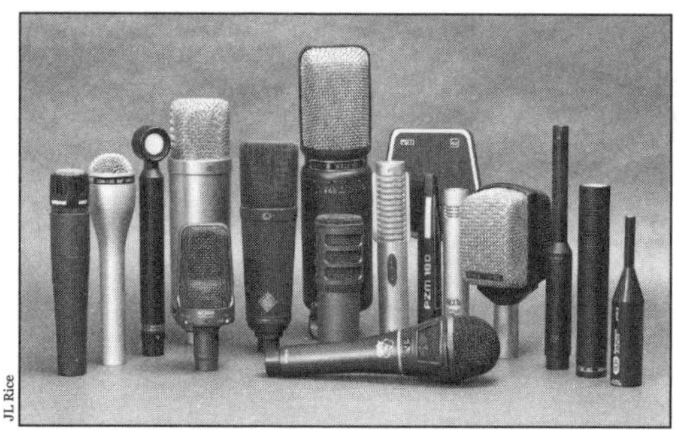

JL Rice

Variety of microphones

Mike Howes, a microphone engineer for Logitech says, “Creativity and teamwork are what I like most about my job. I work as part of a large design team. We sit down together during the initial concept phase of the design and brainstorm. Then we work together to realize that concept into a product that you can see on the shelves at Best Buy. I love being involved in that process and having the opportunity to work with and sometimes create new cutting edge technology.”

Developing a microphone at Holophone

Following is an interview with Michael Godfrey, the inventor of Holophone, a Surround Sound Microphone.

CB: How did you get started? Where did you go to school?

MG: I've been a musician for many years. I got into music on the other side of the glass, playing music for eons. I got into engineering because when I tried to explain to the engineers how I wanted my sound to be, I didn't have the language necessary to do that. So I went to learn how to speak to engineers... thus becoming one myself. So I combined both disciples and all of a sudden, I could produce things because I understood the artists and understood the engineering side of things.

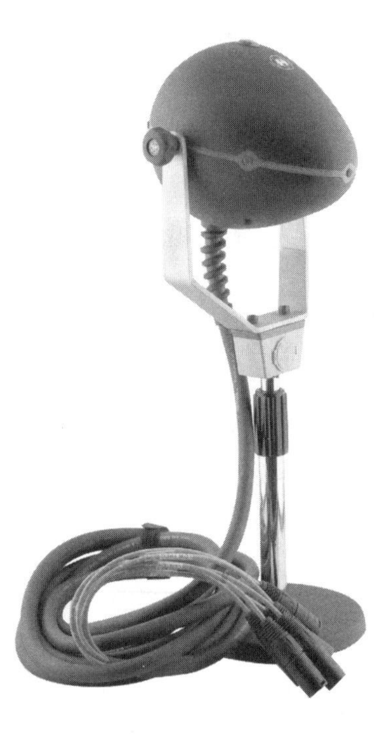

That turned into this company because I took what I wanted to do musically and I had to create a tool in order to do that. The result of that creation process was the 3-D microphones that we sell now. I used the knowledge that I gained in school in recording. I learned how sound works and physics, and I took what I wanted to do as a musician and applied the stuff I learned, and came up with this 3-dimensional microphone. It's specifically designed to take the cost and complexity out of the equation for capturing surround sound, and to simply and instantly provide a sonic, three-dimensional feeling of "being there" as the end result to the viewer.

CB: What is the advantage of a 3-D microphone?

MG: This microphone is specifically designed for 5.1-channel and 7.1-channel surround sound. It takes all the guesswork out of how you make surround sound by actually doing it all for you physically, with no processing, just physics. It emulates the human head and has capsules placed on the human head in correlation to the channels in 5.1-channel playback systems. So when sound is perceived as coming from one direction, it bends around the embodiment, and is picked up slightly from all of them in a way that our brains figure out, on playback, that we are hearing things in three dimensions. And it works! It's been around for 13 years and is sold all over the world.

CB: Sounds like you are growing.

It's not an easy ride. But if you love it and you are doing something that's really good, and you persevere, it will work.

CB: Do you hire engineers?

We have hired engineers and we subcontract. We have predominantly become an equipment supplier. We are a manufacturer. We have lots of people all over the world doing lots of different projects from the Grammys to the Superbowl to the largest events on the planet. We are in touch with a lot of engineers. We listen to the needs and requirements of a lot of engineers. Through my earlier education, I'm able to apply that to creating better products for people in the field and that's the role I'm now playing. We're always working on new stuff.

CB: What advice would you give to a student that wants to do what you do?

If a student likes music and wants to get into microphone design, I would recommend they start by learning the software side (Pro Tools) or the hardware side (electronics). They should first and foremost understand the physics of sound and how that applies to anything. And you really have to learn the basics before you get into digital programs. I learned all on analog gear and because of that I believe I have a much better understanding of what the digital boxes are trying to do in the first place. I think it's very important for students to learn the basics of sound in an analog fashion or they'll never understand what the boxes are doing. You have to focus. Don't get scattered. There are so many things that you can do in this business. It's good to be versatile. Learn everything, but concentrate on what you are doing and then you become much better at it, and that's the way you really succeed in this business.

CHAPTER FIVE

Computer Science and Software Engineering

Once upon a time, computers could only make a beep sound. All the music and sound that comes out of computers today only exists because engineers thought that music would change the face of computing. The earliest computer games used to vary the pitch, length, and frequency of the beep to give different sensations.

Thankfully, this beep frustration spurred manufacturers to create computer sound cards that could produce audio. Suddenly, you could listen to CDs on your computer! It was the beginning a new era in sound enjoyment. Video games could be played in 3-D or surround sound. People suddenly had the capability to record themselves in a microphone on their computer and hear it played back through the speakers because their computer could convert the sound from analog to digital.

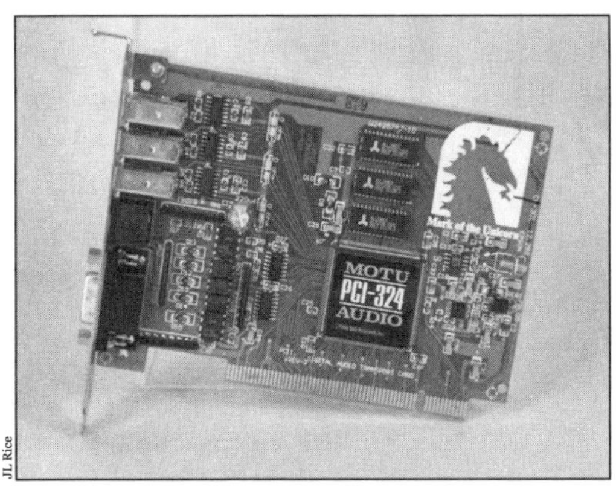

JL Rice

Computer sound card

It was the dawn of multimedia. Without sound on a computer, this wouldn't have been possible.

If a career in sound card design is fascinating to you, get a degree in electrical or audio engineering. You will be well served to have a good understanding of how sound travels, the analog-to-digital conversion when the sound enters the computer, and the digital-to-analog conversion when the sound jumps out of the speakers. Other subjects of interest for this type of engineer are frequency modulation synthesis (the computer overlaps multiple sound waves to make more complex sounds), wave table synthesis (using samples of real instruments to make better recordings), and a good familiarity of surround sound, home theatre systems, and programs like GarageBand or Cakewalk. You will probably be the type of person that loves playing video games and passes countless pleasurable hours downloading music from Napster or iTunes.

Gaming sound engineer

Working in the video game industry as an audio engineer can be a dream come true for the video gamer lover. Audio engineers are so integral to the success of a game they truly bring a game to life. The soundtrack of a game creates and reinforces the mood as well as a more immersive experience for the player. They create the sounds of special effects such as explosions, machine guns, and tires squealing. They may also be responsible for the background music; auditioning and recording the actors; and editing, mixing, and mastering the soundtrack, which will not only be used in the game but may also be sold on CD or for download around the world.

As games get more complex, the demand for audio engineers will only increase. Today, many titles require full orchestras. Rising from the explosion are radio stations such as Games Music

Radio, Video Games Live™ and PLAY! A Video Game Symphony, which features award-winning orchestral music from popular video game titles, including Final Fantasy, Silent Hill, Metal Gear Solid, The Legend of Zelda, Super Mario Brothers, World of Warcraft, Sonic the Hedgehog, and Shenmue.

Outstanding graphics on large screens above the orchestra will accompany the scores to highlight memorable moments from the games. Video Games Live™ is a concert event put on by the video game industry to help encourage and support the culture and art that video games have become. Video Games Live™ bridges a gap for entertainment by exposing new generations of music lovers and fans to the symphonic orchestral experience while also providing a completely new and unique experience for families and non-gamers.

In this growing industry, the engineer must have tremendous musical talent as well as technical proficiency. Some days, the job may require creating spoken instructions, other days the engineer may be creating ambient noise such as rain, a crowded street, or an office building. The sounds may be based on real sounds or they may be completely imaginary for a game that takes place in another world. Other days, the engineer may be composing, scoring, or recording. Experience in recording and mixing is considered highly desirable.

To create the sound design for a game, the engineer works from a creative brief from the programmer. If a sound doesn't exist, the engineer will either improve or modify a previously created sound or create it from scratch. They may have to sync a voice to a talking animated figure, or they may have to record the necessary voice. Although English may be their native language, there is always the possibility that the game will be released in another language, such as French, Spanish, or Japanese.

Gamming sound engineers need excellent communication and teamwork skills. The game designer may have a specific vision that they need to communicate to the audio engineer. The producer may decide to change things at the last minute. The deadlines may be tight and a well-organized audio person will be able to schedule auditions and recordings, and keep up with changes in the schedule or direction of the project. They may also have to keep a record of hundreds of sound effects that were added to the game.

Companies that are hiring this type of engineer will usually look for technical ability but may be more concerned with finding a person who possesses the following traits:

- Is resourceful and energetic
- Work well in teams or independently
- Communicates effectively
- Is imaginative and creative
- Has a good sense of fun
- Has a wide range of music interests
- Loves games
- Works well under pressure
- Has the ability to compose and play music
- Has recording, editing, and mixing experience
- Is detail oriented

Winifred Phillips, the audio engineer for the game "The DaVinci Code" got the job by competing in an audition. "To find a composer to score The Da Vinci Code game, the developer and the publisher conducted a competitive audition, in which composers wrote, recorded, and submitted a new piece of music that demonstrated their unique vision for how the score of the game should sound. I knew that the audition was extremely competitive–some of the top game composers in the industry submitted music for consideration. It was a nerve-wracking experience for me. Given the fantastic story told by The Da Vinci Code and the stellar track record of both the publisher 2K Games and the developer Collective Studios, I knew that this game would be extraordinary.

I really wanted to be a part of it! I read and reread the book over and over, obsessed on the role that music should play in the game, and worked harder on that one piece of music than anything I'd ever written before. When it was done, all the musical elements were there... the clockworks, the classical and contemporary orchestration, the Latin lyrics containing hidden meanings derived from Dan Brown's story... all the building blocks for the final score were in that one track. Later, after I found out that I had won the job, Collective Studios let me know that the music I'd submitted was exactly what they wanted for the game—which was a unique sound, unlike what had been heard in video games before. They told me to go ahead and write the score in the same style I'd used to create the audition track."

If video game sound is your thing, the University of California offers an extension class called Composing Music for Video Games. By reading the course description below, you will get a feel for the challenging nature of this field.

> Course description: Through weekly writing assignments, lectures, audio-visual demonstrations, and visits by guest speakers, composers interested in exploring the challenges of creating music for video games are presented with an overview of the composition techniques, organization, and delivery formats unique to the video game industry. Topics include in-game versus cinematic scoring; budgeting and project management; contracts; technology tools for asset creation and processing; music engines and compositional techniques specific to video game music; electronic music creation versus using live musicians; mixing; composing for different game genres (MMOG, FPS, RTS, educational); and audio formats and delivery of assets for different console formats (SKUs), such as Xbox/Xbox 360, PS2/PS3, PC, GameCube, and handheld devices (GameBoy, Nintendo DS). Weekly composition

assignments focus on writing original adaptive music similar to works created for current games. Participants also have the opportunity to compose MIDI/digital music with Pro Tools (note: this is not a Pro Tools course; Pro Tools are available for use in the class but students may compose with any tools of their choice). For guest speaker information visit uclaextension.edu. Prerequisite: Some composition and theory background. As Pro Tools is only available in-class and outside composition projects are assigned, students must create final mixes with composition tools they have access to outside of class, such as Pro Tools, Logic, Digital Performer, Cubase SX, Nuendo, Sonar, or Reason.

Ringtone engineer

JL Rice

Ringtone engineers are sometimes referred to as audio designers and often work in research and development (R&D) for wireless device companies such as Nokia and Verizon. You may integrate audio hardware and acoustics or audio modules for mobile phones or handheld multifunctional devices, or you may find yourself working on electro-acoustics, measurement, or hardware design.

This field is growing rapidly and requires at least a bachelor's degree in electrical or audio engineering or computer science to get your foot in the door. Positions in this filed require a self-starter that can handle the pressure of deadlines. As more people purchase handheld devices, the need for excellent ringtone engineers will only increase.

CHAPTER SIX

Digital Music

We can all agree that digital music has changed the way we listen to and enjoy music. First, music was put on CD in 1983 when Sony and Phillips introduced the first CD. That was fun because the small size made it more portable and the sound was better than the average cassette tape walkman. Selecting a song with a button was so much better than fast forwarding or rewinding a tape. We felt liberated. After a while, we even came to expect the ease of a push button. If we wanted to buy a cassette tape or album in the store, however, the selection was dwindling as the entire recording industry moved to a digital format.

JL Rice

Recording software

The next logical step in digital music was the MP3 player. I mean, who wants to carry around a stack of CD's and a CD player anyway? These new portable MP3 players were great because they were pocket-size. The only problem was that when they first came out, they could only hold about 10 songs or 1 CD. Still, we were excited to have something so small and powerful. We could take our favorite songs off of any CD we owned and take them with us. These first players were such a hit that we can now put entire collections of music on them. The memory has jumped from 10 to 15,000 songs—plus photos and video.

Digital music has changed the way we, as a society, listen to and enjoy music. It has changed the way that music is recorded, mixed, and produced. Engineers that work in this digital music player industry must have a desire to always be on the cutting edge. They must be life-long learners because this industry changes quickly. If you are not really interested in music and the transportation of your soul as a result of listening to something really moving, you may get left behind.

iPod engineers

This book would not be complete without a look at the iPod Engineering division of Apple Computers. iPod took the world by storm. The release of their first iPod had people clamoring to get one because it could hold more music than any other MP3 player on the market. The white "ear buds" were a statement that you were among the hip and trendy. You were a person that wanted to be a part of a heavily anticipated wave of the future.

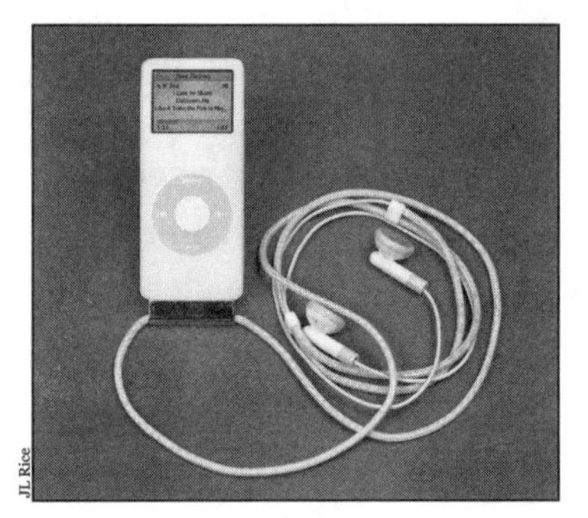

JL Rice

iPod nano

People started cults and dressed as iPods for Halloween. Friends walking dogs would even stop, swap ear buds and listened to each other's music for a few minutes. The iPod was no ordinary music player and the engineers of the iPod engineering division are not ordinary engineers. Kate, of Apple's iPod platform technology division says, "There's really nothing quite like working on a product that almost everyone covets. It's a great product line and a terrific team; somehow we manage to keep topping ourselves. It's a lot of hard work, but it's a really professional team, and that makes a big difference."

Imagine for a minute that you are an engineer at iPod and while taking the bus home from work, you see people getting on or off that have the white ear buds dangling. How would it feel to know that you were a part of the product? You were a part of the revolution. You were a part of the enjoyment or escape these people have everyday. Alex, of the iPod engineering division, says, "Every day, we have the opportunity of being involved in owning decisions that impact the future of the incredible products we create. It makes us proud to walk around anywhere we go and see the fruits of our labor. Those white ear buds are everywhere!"

Do you want to be an iPod engineer? It takes much more than technical skills to be this type of engineer. The iPod engineering division employs about 20 or more people to produce the music player. To have any finished product, these people have to work together as a team. According to Apple's website, the iPod engineering division encompasses a variety of disciplines, including: Software Engineering, Software Architecture, Electrical Engineering, Hardware Engineering, Industrial Design, Product Design, Project Management, and Quality Assurance.

MP3 players

MP3, or MPEG (Moving Picture Experts Group) Audio Layer 3, is a method of compressing video and audio files. It has been around since the late 1990's and enables music to be compressed while still retaining near-CD quality sound. The Fraunhofer Institute in Germany patented the MP3 format in 1989. Not only can MP3s be song files, they can also be audio recordings of podcasts or your own voice or band.

Think of it like this—a song on a CD uses about 10 MB per minute. A song that is three minutes long requires about 30 MB of space. A program such as iTunes that can rip and encode (compress) the song to MP3 has the ability to reduce the size requirement from 30 MB to 3 MB. As a result, getting a 1GB memory card for your player will enable you to carry about 20 CDs. Essentially, by purchasing several memory cards (SmartMedia, Compact Flash, etc.), you can take your entire collection with you anywhere and it will only weigh four to six ounces! This technology is changing the face of how we collect, access, listen to, and store music and audio files.

MP3 is an amazing format because of the design. Engineers figured out how to compress a song or any audio file while retaining a majority of the original sound by using a technique called perceptual coding. Perceptual coding can improve the representation of digital audio while reducing the file size at the same time. For example, if two songs are playing at the same time, we hear the louder one. Anyone who has blown a dog whistle knows that the human ear cannot hear everything. Some sounds come through better than others. Using these theories, MP3 compresses songs by eliminating what most people can't hear anyway. As a result, it fits into a smaller space and for a portable device, still sounds great.

Other file formats supported by most MP3 players include: Windows Media Audio (WMA), Waveform Audio (WAV), Music Instrument Digital Interface (MIDI), Advanced Audio Coding (AAC), Advanced Streaming Format (ASF), and Vector Quantization Format (VQF).

MP3 players contain itty-bitty hard drives or they use solid state memory. If the player uses solid state, there are no moving parts, and it acts like a memory or storage device. The software inside the player allows you to organize, transfer files to the player, and play songs at will. When you hit the "play" button the player pulls the song from its internal file cabinet, decompresses the MP3 file, converts it from digital to analog (this is necessary to create sound waves) and amplifies it so you can hear the song through speakers or headphones. Voila! It sounds so simple, but the engineers behind the technology might tell you differently. If the player contains a tiny hard drive, you can transport pictures, video, and files on you player just like any portable hard drive.

If you like working with sound files, writing software, computer programming, analyzing compression techniques, and understanding how we hear sounds, this may be a good fit for you. This type of job requires imaginative solutions to problems and the ability to work well with other contributors of the project.

CD and DVD manufacturing

Another industry in the music sector that employs many engineers is in production and manufacturing of CDs and DVDs. Engineers trying to break into the industry often overlook this area.

A CD usually holds between 650 and 783 MB. They are about 1.2 mm thick and made of plastic with a thin layer of reflective aluminum on top. Microscopic pits or bumps assembled into a long chain are laid into the plastic starting with the inside and working out. The laser reads the series of bumps by reflecting light off of the aluminum layer and detecting the changes in the amount of light. The "brain" in the player processes these changes and sends the message to the speakers. The technology is really amazing when you think about the song being sent as a series of 1's and 0's, converted to sound waves, and played by your speakers in a matter of seconds.

To read a CD, the CD player motor must spin the CD past the laser head that is reading the bumps or pits. The laser is attached to a tracking mechanism that follows the spiral track of bumps, which, if laid out in a straight line, would be as long as 3.5 miles! Another cool aspect of the tracking device is that when the laser is reading at the beginning or center of the disk, the speed of the rotation must slow down. If you are in a running race around a track, do you want to be on the inside lane or outside? The inside lane allows you to go faster and you will probably win the race. The same is true with CD tracking, except the goal is not to win the race, the goal is to keep the same pace throughout the race.

Optical CD/DVD drive rack

Engineers that work in the CD business are developing new methods of encoding, writing new software programs to enhance our listening enjoyment, developing readers that may read right through dust or scratches, developing new error correction processes, streamlining manufacturing processes, and searching for new materials to make CDs smaller, more efficient and/or increase capacity.

CHAPTER SEVEN

Other Careers

Industrial audio

While sound engineering is mostly associated with music recording and live event sound reinforcement, industrial audio is another category that is often overlooked. Several sub-categories of industrial audio exist, such as:

- Installed sound. This category comprises the design and installation of sound systems for restaurants, coffee houses, airports, stadiums, theaters, performance spaces, etc.
- Telecom. Examples of Telecom include on-hold music, on-hold advertising, and phone tree voice-over recordings.
- Device audio. These consist of audible prompting systems and mobile recording systems.

Portable defibrillator audio

By Robert Smith – Principle Scientist, Acoustic Systems, Physio-Control

Specifically, I work for a major medical electronics manufacturer as an acoustic researcher. The devices we make are defibrillators and biological monitors carried by paramedics and first responders. There is a specific category of defibrillators called automatic external defibrillators (AED). These are designed to assist first responders and lay rescuers in performing resuscitation during an acute cardiac event. A critical component of the user interface for this class of devices is the audible prompting system or audio instructions that aid the rescuer by suggesting treatment such as "Check for pulse", "Shock advised", "Start CPR", etc. The audible prompts are recordings of voice actors and actresses in over 30 languages. Prompt scripts are designed to rapidly and clearly communicate information. Additionally, many AEDs are equipped with an on-scene audio recording system to capture sounds and dialog during resuscitation.

JL Rice

Robert Smith's audio studio

The following major topics are necessary to achieve a high level of audio performance for medical devices:

1. Audio playback system – Designing a portable AED audio playback system is exceptionally challenging given the many conflicting goals, agency requirements (IEC, UL, TUV, etc.) and manufacturing constraints. The audio should have a high degree of speech intelligibility to be heard in noisy environments. Loudness is only one aspect of this goal. The system is necessarily constrained in power by the portable, battery powered nature of the device. The speaker implementation with the case must both consider speaker acoustics and environmental concerns such as preventing the intrusion of fluids into the case. Target manufacturing costs may require the use of a lower sample rate system than found in consumer audio.

2. Audio recording system – The intention of the AED on-scene audio recording is to capture dialog and sounds occurring during an acute cardiac event. The device itself is periodically producing very loud prompts, the environment may be noisy, such as in an airport concourse or a stadium event and many people will be talking simultaneously. On the other extreme, the AED may be recording in a quiet bedroom with hardly any sounds other than the voice prompts occurring. Again, the system must be designed with attention to conflicting goals, agency requirements and manufacturing constraints.
3. Speech intelligibility research – As can be seen from the brief descriptions for audio recording and audio playback systems above, speech intelligibility is an extremely important topic for an emergency medical device that incorporates them. Research in this area includes perceptual acoustics, noise masking, reverberant environments, and vehicle interior noise, just to name a few.
4. Voice prompt script design and translations – Every message an AED utters is scrutinized and reviewed carefully to create a prompt that clearly communicates while minimizing the possibility of confusion or misunderstanding. The prompt should have a minimum of words that can rapidly convey the message. This is then repeated for localizations in the various languages/ localizations supported.
5. Recording voice actors and actresses – Once the voice prompt scripts and translations have been completed, they need to be recorded by a special class of professional talent known as voice actors and actresses. Voice-over work requires extensive training to achieve good diction, steady delivery cadences, clear enunciations, and proper

voice inflections. Attention to cultural issues is of extreme importance to avoid producing an offensive product.

6. Voice prompt processing – When a recording of a musical group has been mixed, there is another stage called mastering. One of the goals in mastering is to have the music translate well to the consumers' media playback systems to provide a positive acoustic experience. The qualities, acoustic environments, performance levels for the playback systems vary considerably. An AED, on the other hand, is a very specific playback system. Using this knowledge, the recorded voice prompts are processed in very specific ways to maximize performance for each product.

Forensic audio engineer

If you like "who dunnit" mysteries and want a career in criminal investigations without wearing blue or stepping into the line of fire, forensic audio may be a good choice. Forensic audio is the application of audio knowledge, techniques and methodologies to solve crimes or public debates and discussions. The word "forensic" is defined as "Pertaining to, connected with, or used in courts of law or public discussion and debate." (The Random House Dictionary of the English Language, Second Edition - Unabridged. Random House, Inc.,1987).

Audio forensics were used to solve mysteries in the John F. Kennedy assassination. When researchers wanted to know how many gun shots there were, the time interval between shots, and whether one of the shots was an echo, they turned to forensic audio specialists to get the answers. When the Watergate tapes were discovered, it was the forensic audio scientist who determined that some of the tape had been erased and that there was an 18-minute gap in the tape.

If forensic audio is interesting to you, you will be using sophisticated technologies and scientific methods to aid historical, criminal, and civil investigation. Some of the specialty areas include:

- Voice identification
- Audibility analysis
- Enhancement
- Authenticity analysis
- Sound identification
- Event sequence analysis
- Dialogue decoding
- Other signal analysis

Instrument and technology sales

When you consider the large range of music equipment that you've read about in this book, it shouldn't be hard to see that every manufacturer also needs technical sales people to sell and sometimes install the equipment. This engineer or technician may also provide phone support.

JL Rice

Celeste Baine and John Hardy discuss design.

Anyone that plays, appreciates, or produces music is a much better fit for instrument sales than someone who is just a great salesperson. Instruments are used by musicians, schools, music companies, studios, and more. Gino Sigismondi, an applications engineer for Shure, Inc. says, "I help people choose and use Shure products. I help them figure out the proper product for their application, how to make it work, how to hook it up, and I help them troubleshoot it if something goes wrong." Gino received a music business degree (a music major with a business minor). In that degree program, there are certain classes that are geared toward the music business, such as music publishing, concert promoting, recording engineering, and live sound reinforcement applications.

Noise control design/acousticians (airports, concert halls, schools, restaurants, car interiors, highways)

Noise control is usually a multidisciplinary degree that will be offered through a college of engineering. Students may enroll in audio engineering, mechanical engineering, or aerospace engineering if they are interested in vibrations or noise control. Topics covered will usually include vibration isolation, energy absorption, properties of materials, sound barriers, dampening systems, and solutions. This degree program can be very versatile because noise control is an important issue for many industries.

College/university music or audio technology teacher

Every audio engineering program needs instructors. The role of this professional in the college or university is highly diversified. Depending on the specialty, they may be teaching anything from GarageBand, to audio for video production, to console design. Within a school of music, there are teachers of performance, theory, composition, history, and education. They may be called upon to work the technical end of a performance or composition. They may find themselves researching equipment or calling manufacturers for demos or new gear. These instructors may further specialize in a certain style of music or a music environment, such as church music, music therapy, or commercial music.

CHAPTER EIGHT

Getting Started

When you work with other people, a very important skill is communication. Good communication is the ability to listen, write well and speak well. In addition, these engineers need to be passionate, energetic, focused, and excited about their work. The music industry is always evolving and changing. Do you think that an industry that strives to be on the cutting edge wants to hire engineers that are not cutting edge types of people? Being a good engineer is so much more than being a brain. Take extra classes in technical writing, develop excellent presentation skills, choose an industry that makes you excited about your career, and commit yourself to being the best you can be. Four years in engineering school may not be enough. Focus your energy, meet your goals, pay attention to details and push yourself to be creative. If you want to be a champion, you have to start acting like one.

Remember that this book is only one source of information to help you decide whether you want to become an engineer. Right now, you need to begin reading everything you can find about engineering and talk to every engineer or engineering student you know about the challenges ahead and how to prepare for them. Attend a summer camp or program pertaining to engineering at your school. Obtaining this information now may save you lots of heartache if you decide later that you are on the wrong path.

Academic preparation is also essential to exploring engineering as a career. In addition, getting involved in extracurricular activities pertaining to engineering can give you invaluable exposure. In high school, classes in algebra I and II, trigonometry, biology, physics, calculus, chemistry, computer programming, or computer applications can tell you if you have the aptitude and determination to study engineering. All of the above courses are not required to get into every engineering school, but early preparation can mean the difference between spending four years in college or six. Some universities also require two to three classes in a foreign language for admission. Check into the programs that interest you and begin to fulfill their requirements. Advance placement or honors courses and an ACT score of 20 or SAT of 1000 are recommended.

Mark French, the professor that developed the Mechanical Engineering Technology Acoustics Lab at Purdue University advises students that want to work in this industry to know:

- How to estimate numbers
- How to talk to musicians – how to translate 'dudespeak' into rigorous engineering terms
- How to clearly explain sound and vibration concepts to non-technical people
- Some music theory
- How to play at least one instrument passably

Gino Sigismondi, an applications engineer for Shure, Inc. says, "If you want to work for an audio manufacturer, the most valuable thing you can do is to get practical experience. Get out and actually do the work. Take the recording and computer courses in college and learn the basics from a technical standpoint, because you really need to know that, and unfortunately, a lot of programs don't teach it. They might teach you how to set up mics and record a band and how to turn the knobs, but it's important to know why are you turning that knob, and when you turn it, what's going on under the hood. Try to learn more about the technical side. When someone says that you are using a low impedance microphone, what does that mean? Ask what's impedance? Why is low impedance better than high impedance? What numbers go with that? Try to really get down to the nuts and bolts of what you are learning and then get out and do it. If you go to work for a manufacturer, that experience will be invaluable because you'll be able to relate to the customer of the products and talk on their level. And it's also invaluable for the company, because you bring hands-on, practical information back to the company from actually knowing how to use the products that the company makes. The hands-on portion is really important."

Junior Engineering Technical Society (JETS)

JETS is a national society dedicated to providing students with guidance and information about engineering. The Society offers many programs that will help you decide if engineering is the career for you. Teachers and parents also gain exposure to the social, political, and economic impact that engineering can have on our lives.

If JETS interests you, check out their Web site www.jets.org. The site is packed with additional information about the different careers in engineering, programs offered, competitions, activities, and events. JETS is an excellent source of information for you as you continue to explore the many aspects of engineering as a career.

Summer camps

Summer camps provide another innovative approach to preparing for a career in engineering or evaluating if that career is right for you. Find out what it is like to study engineering, about the different types of engineers, and what engineers do on a daily basis. Almost every college of engineering offers a residential or commuter summer engineering camp for high school students. They offer students a week or two of fun while developing leadership and professional and personal organizational skills, and they provide opportunities to meet and talk with engineers during visits to local engineering companies. Check with the college of engineering at a university near you to see if any summer programs are offered, or visit the Engineering Education Service Center's Web site at www.engineeringedu.com to find a camp in your area.

Student competitions

A great way to get a feel for engineering is to look at the student design competitions that are sponsored or co-sponsored by various engineering societies and organizations. These competitions are developed to encourage and motivate students. They focus on teamwork and allow the students to get a "real-world" look at the design process, cost of materials, team dynamics, and environment.

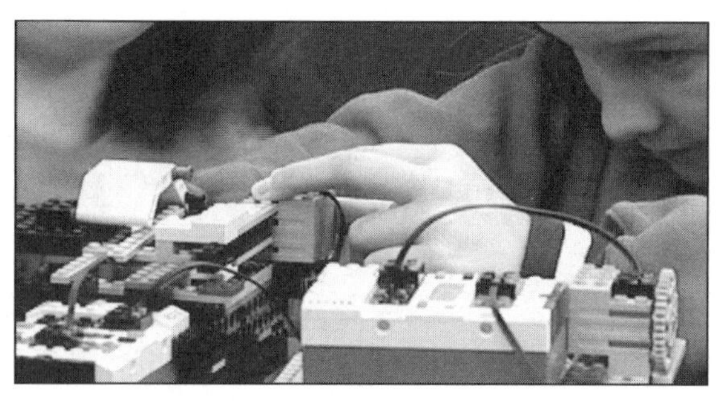

Student builds a LEGO® model for competition.

A few of the more popular competitions include:

- Boosting Engineering, Science, and Technology (BEST). This robotic competition provides students with an intense, hands-on, real engineering and problem-solving experience that is also fun. www.bestinc.org

- FIRST Robotics Competition. Corporations and universities team up with high schools in a high-tech, robot sporting event. www.usfirst.org
- Mathcounts. This is a national math coaching and competition program for 7th and 8th grade students. www.mathcounts.org
- Future City. Students learn about math and science in a challenging and interesting way through reality-based education using SimCity software. www.futurecity.org
- TEAMS and NEDC. These events are sponsored by JETS. www.jets.org
- International Bridge Building Contest. For information, visit www.iit.edu/~hsbridge
- Odyssey of the Mind. This world-wide program promotes creative team-based problem solving for kids from kindergarten through college. www.odysseyofthemind.com
- National Science Bowl. This is the U.S. Department of Energy's academic competition in which teams of high school students answer questions on scientific topics in astronomy, biology, chemistry, mathematics, physics, earth, computer and general science. www.scied.science.doe.gov/nsb
- BattleBotsIQ.– This is a comprehensive educational program where students learn about the science of engineering through robot building. www.battlebotsiq.com

The AES also has an audio competition for undergraduates. For a more comprehensive list of competitions and descriptions, visit the Engineering Education Service Center's Web site: www.engineeringedu.com.

Appendix

Glossary

Music, like any subject, has its own terminology. This section will introduce you to the terms used most often to talk about music.

Acoustics - The study of sound and its behavior within an environment.

AES - Acronym of the Audio Engineering Society.

Amplifier - A device that increases the amplitude, power or current of a signal. The resulting signal is a reproduction of the input signal as well as this increase.

Analog - An analog audio signal is represented by variations such as voltage speed or frequency and the strength of amplitude or volume of an electrical audio signal. The audio outputs from a computer's soundcard or synthesizer are typically analog outputs even though the file being played is digital through a digital to audio converter.

Audio Frequency - Signals in the human audio range: nominally 20Hz to 20kHz.

Audiophile - A person enthusiastic about sound reproduction who is discerning about the quality of the audio.

Band Pass Filter (BPF) - Filter that removes or attenuates frequencies above and below the frequency at which it is set. Frequencies within the band are emphasized. BPFs are often used in synthesizers as tone shaping elements. A filter which allows only certain audio frequencies to pass, while rejecting all others above and below the cutoff points. An example of a BPF may be found in a "3 - way" loudspeaker system that will utilize a "woofer" for bass frequencies, a "midrange" unit for middle frequencies, and a "tweeter" for high frequencies. Whilst the woofer (which has no frequencies below it) will be able to have it's band of frequencies fed to it via a "low pass" filter and the "tweeter" which has no frequencies above it will have a high pass filter, the midrange, which will have frequencies both above and below it's area of operation will need to have its frequencies fed to it via a BPF.

Bi-Directional microphone - Microphone that will pick up sounds that are emanating from the front of the microphone (on-axis), and the rear of the microphone (off-axis), and largely reject those to the side. Also described as a "figure of eight" microphone, as the field pattern just described corresponds to that of the figure eight.

Breadboard - A breadboard is a reusable solderless device used to build a (generally temporary) prototype of an electronic circuit and for experimenting with circuit designs.

Cakewalk - Cakewalk is a developer of powerful and easy to use products for music creation and recording. These products include digital audio workstations and sequencers, music software and virtual instruments.

Capacitor Microphone - Microphone that operates on the principle of measuring the change in electrical charge across a capacitor where one of the electrodes is a thin conductive membrane that flexes in response to sound pressure.

Condenser Microphone - A microphone that generates an electrical signal when sound waves vary the spacing between two charged surfaces, specifically the diaphragm and the back plate. Condenser microphones offer the greatest fidelity in terms of traducing sound waves into an electrical signal, however, they do have disadvantages such as picking up hums/ ground loops etc, and a delicacy which renders many of them more suitable for studio, rather than stage use.

Cooperative Education (Co-op) – Program that combines real-world experience with college classes.

dB (Decibel) - A logarithmic measure of sound pressure level. Someone with pretty good hearing will be able to pick up sounds from 0 - 10dB, a quiet room can be as much as 40dB, and people will start to feel pain and possibly sustain hearing damage if they are exposed to levels in excess of 135dB for any length of time (The public address speakers of the kind used at very large concerts may realize this if they are running at their maximum.)

Digital Audio - The numeric representation of sound. Typically used as the means for storing sound information in a computer or sampler.

Drum Machine - An electronic device, usually controllable via MIDI commands, that contains samples of acoustic drum sounds. Used to create percussion parts and patterns. EQ (Equalization) - Sophisticated tone controls that can subtly enhance or drastically change a sound. Can make a sound "brighter", "darker", "fuller" or "tinny." Controls include selecting frequency to be affected, bandwidth and amount of boost and cut.

Eno, Brian - Perhaps best known as a musician and producer, he's also an artist, professor and thinker. Music-wise, even if you haven't heard any of his own records, you may have heard his production contributions to albums by rock legends U2, David Bowie or James. In other media, his music sometimes appears in films, television programs and commercials, and the Windows 95 start-up sound.

GarageBand - GarageBand is a software application that allows users to create a piece of music. It was developed by Apple Computer for their Macintosh computers.

High Pass Filter - Utilizes capacitors, as the impedance of a capacitor decreases for High frequencies, this enables it to ensure the passage of higher frequencies, and stop the passage of more powerful lower frequency signals.

MIDI - Acronym for musical instrument digital interface; technology standard that allows networking of computers with electronic musical instruments.

MIDI Channel - An information pathway through which MIDI information is sent. MIDI provides for 16 available channels, each of which can address one MIDI instrument.

Mixer - A hardware or software device that combines multiple audio signals into one destination signal. Mixers usually provide control over the volume and/or stereo balance of each source signal and are used as a recording device that allows several different audio sources to be combined. Provides independent control over each signal's loudness and stereo position.

MP3 (.mp3) - MPEG Layer III, digital audio compression format achieving smaller file sizes by eliminating sounds the human ear can't hear or doesn't easily pick up. It is an encoding format that takes out all the irrelevant data in a recording and compresses the remaining data. An MP3 file can be 1/12 the size of an original recording taking up far less space on a computer's hard drive, making it feasible to email the audio file, post on the web, make MP3 CDs and use with personal music player.

Musical - Genre of twentieth century musical theater, especially popular in the United States and Great Britain; characterized by spoken dialogue, dramatic plot interspersed with songs, ensemble numbers and dancing.

Orchestra - A performing group of diverse instruments; in Western art music, an ensemble of multiple string parts with various woodwind, brass and percussion instruments.

Post Production - Work done to a stereo recording after mixing is complete.

Pro Tools - Pro Tools is a digital audio workstation by Digidesign for music production and digital audio editing. It is widely used to create audio for film, television, and music.

Rip - to extract or copy data from one format to another more useful format. The most common example is found in the phrase "CD Ripping" which means to copy audio tracks from an ordinary audio CD and save them to hard disk as a WAV, MP3 or other audio file, which can then be played, edited or written back to another CD.

Sweet Spot - The optimum position for a listener within the sound field created by a pair of stereo speakers, or the optimum position for a microphone relative to it's pickup pattern and the sound field created by whatever is being recorded. The optimum position for a microphone, or for a listener relative to monitor loudspeakers.

Synthesizer - Electronic instrument that produces a wide variety of sounds by combining sound generators and sound modifiers in one package with a unified control system.

Industry Contact Information

For additional career and education information, contact the following:

Audio Engineering Society
International Headquarters
60 East 42nd Street, Room 2520
New York, New York 10165-2520, USA
Ph. 212 661 8528, Fax 212 682 0477

Academy of Motion Picture Arts and Sciences
8949 Wilshire Boulevard
Beverly Hills, CA 90211, USA
Ph. 310 247-3000, Fax 310 859-9351

Acoustical Society of America (ASA)
500 Sunnyside Boulevard
Woodbury, NY 11797, USA
Ph. 516 576-2200, Fax 516-349-7669

American National Standards Institute (ANSI)
11 West 42nd Street
New York, NY 10036, USA
Ph. 212 642-4900, Fax 212-302-1286

Association of British Theatre Technicians
55 Farringdon Road
London
EC1M 3JB
Tel: 020 7242 9200

Association of Professional Recording Studios (APRS)
2 Windsor Square, Silver Street
Redding, RGl 2TH Berkshire, UK
Ph. +44 734 756-218, Fax +44 734 756-216

Billboard Magazine
1515 Broadway
New York, NY 10036, USA
Ph. 212 764-7300, Fax 212 536-5358

Broadcast Education Association (BEA)
1771 N Street, N.W.
Washington, DC 20036, USA
Ph. 202-429-5355

Broadcasting Entertainment Cinematograph and Theatre Union (BECTU)
373-377 Clapham Road
London
SW9 9BT

Canadian Recording Industry Association
1250 Bay Street, Suite 400
Toronto, Canada MSR 2B1
Ph. 416 967-7272, Fax 416 967-9415

Electronic Industries Association (EIA)
2001 Pennsylvania Avenue N.W.
Washington, DC 20006-1813, USA
Ph. 202 457-4900, Fax 202-457-4985

European Broadcast Union (EBU)
Ancienne Route 17A, Case Postale 67
CH-1218 Grand Saconnex, Geneva, Switzerland
Ph. +41 22 7172111, Fax +41 22 7172481

Institute of Electrical and Electronic Engineers (IEEE)
345 East 47th Street
New York, NY 10017, USA
Ph. 212 705-7900, Fax 908 463-3657

International Alliance of Theatrical and Stage Employees (IATSE)
1515 Broadway, Suite 601
New York, NY 10036, USA
Ph. 212 730-1770, Fax 212 921-7699

Intertec Publishing Corp.
9800 Metcalf
Overland Park, KS 66212-2215, USA
Ph. 913 341-1300, Fax 913 967-1898

Linda Hall Library of Science, Engineering & Technology (formerly the "Engineering Societies Library" in New York)
5109 Cherry Street
Kansas City, MO 64110, USA
Ph. 816 363-4600, Fax 816 926-8790

Mix Magazine
6400 Hollis Street, #12
Emeryville, CA 94608, USA
Ph. 510 653-3307, Fax 510-653-5142

Music and Entertainment Industry Educators Association (MEIEA)
David Hibbard
McLennan Community College
1400 College Drive
Waco, TX 76708, USA
Ph. 817 750-3578

National Academy of Recording Arts and Sciences (NARAS)
3402 Pico Boulevard
Santa Monica, CA 90405, USA
Ph. 310 392-3777, Fax 310 392-2778

National Association of Broadcasters (NAB)
1771 N Street, N.W.
Washington, DC 20036, USA
Ph. 202 429-5300, Fax 202 775-3520

National Association of Music Merchants (NAMM)
Education Department
5140 Avenida Encinas
Carlsbad, CA 92008, USA
Ph. 619 438-8001, Fax 619 438-7327

National Society of Professional Engineers (NSPE)
1420 King Street
Alexandria, VA 22314, USA
Ph. 703 684-2800, Fax 703 836-4875

Production Services Association
PO Box 2709
Bath

BA1 3YS
Tel:01225 332668

Professional Lighting and Sound Association
38 St Leonards Road
Eastbourne
East Sussex
BN21 3UT
Tel: 01323 410335

Pro Sound News
2 Park Avenue
New York, NY 10016, USA
Ph. 212 213-3444, Fax 212 213-3484

Pro Sound News Europe
8 Montague Close, London Bridge
London, England SE1 9UR
Ph. +41 0207 9408545, Fax +41 0207 4077102

Recording Industry Association (RIAA)
1020 19th Street N.W., Suite 200
Washington, DC 20036, USA
Ph. 202 775-0101, Fax 202 775-7253

Society of Broadcast Engineers (SBE)
8445 Keystone Crossing, Suite 140
Indianapolis, IN 46240, USA
Ph. 317 253-1640, Fax 317 253-0418

Society of Motion Picture and Television Engineers (SMPTE)
595 West Hartsdale Avenue
White Plains, NY 10607, USA
Ph. 914 761-1100, Fax 914 761-3115

SPARS: Society of Professional Audio Recording Services
9 Music Square South, Suite 222
Nashville, TN 37203, USA
Ph. 800 771-7727, Fax 615 846-5123

School Directory

This directory should not be used as your sole source of educational programs. This directory only lists associate's, bachelor's and master's degree programs. It doesn't cover the many schools that offer courses that are not normally in an accredited qualification or certificate. It does not cover the courses that result in an in-house certificate/diploma (e.g. privately-run, non-accredited training courses, product courses). It also doesn't cover courses resulting in an accredited vocational qualification or certificate.

Associate's Degree Programs

Art Institute of New York City
New York, NY 10013
ainyc.aii.edu
Video Production - Associate's of Occupational Studies

Art Institute of Seattle
Seattle, WA 98121
www.ais.artinstitutes.edu
Audio Production - A.A.A.

Brown Institute
Minneapolis, MN 55407
Electronics Technology - Associate's of Electronics Technology

Cayuga Community College (SUNY)
Auburn, NY 13021
www.cayuga-cc.edu
Audio Production - A.A.S.

Cedar Valley College
Lancaster, TX 75134
www.dcccd.edu/cvc/director/la/music/music.htm
Recording Technology - Associate's Degree

Finger Lakes Community College
Canandaigua, NY 14424
Music Recording - A.S.

Full Sail Real World Education
Winter Park, FL 32792
www.fullsail.com
Computer Animation - A.S.
Digital Media - A.S.
Film and Video Production - A.A.
Game Design and Development - A.S.
Recording Arts - A.S.
Show Production and Touring - A.S.

Globe Institute of Recording and Production
San Francisco, CA 94107
www.GlobeRecording.com
Audio Producer - Associate's Degree

Houston Community College System, Northwest College
Houston, TX 77043
Audio Recording - A.A.S.

Indiana University School of Music
Bloomington, IN 47405
www.music.indiana.edu/department/audio
Audio Technology - A.S.

Kansas City Kansas Community College
Kansas City, KS 66112
www.kckcc.edu/music
Audio Engineering - A.A.S.
Music Technology - Associate's of General Studies with an emphasis in Music Technology

Long Beach City College
Long Beach, CA 90808
www.lbcc.edu
Commercial Music: Music Technology - A.A.
Commercial Music: Record Producer - A.A.
Commercial Music: Recording Engineering - A.A.
Commercial Music: Recording Engineering - Certificate

Los Medanos College
Pittsburgh, CA 94565-5197
Music/Recording Arts - A.A. or Certificate of Completion

Madison Media Institute
Madison, WI 53718
www.madisonmedia.com
Recording and Music Technology - Occupational A.A. - Recording and Music Technology

McNally Smith College of Music
St. Paul, MN 55101
www.mcnallysmith.edu
A.A.S. in Music, Emphasis: Music Production – AAS in Music Recording Technology - A.A.S.

Miami-Dade Community College, School of Film and Video
Miami, FL 33167
Broadcasting - A.A.
Film Production - A.S.
Radio and Television Production - A.S.

Mt San Jacinto College
San Jacinto, CA 92583
www.msjc.edu/music/audiotech/
Studio recording: Beg, inter. adv. – Certificate or degree

Northeast Community College
Norfolk, NE 68701
www.northeastaudio.org/
Audio/Recording Technology - A.A.S.

Northwest College
Powell, WY 82435
www.northwestcollege.edu
Music Technology/Multimedia - A.A.S.

Red Wing/Winona Technical College
Red Wing, MN 55066
Electronic Music Technician - A.A.S.

Santa Fe Community College
Santa Fe, NM 87505-4187
Basic Multi-Track Recording - component of certificate and associate's
Digital Audio - component of certificate and associate's
Intermediate Multi-Track Recording - component of certificate and associate's

Shoreline Community College
Seattle, WA 98133-5696
success.shoreline.edu/sccstudio
Audio Engineering - A.A.S.
Digital Audio Production - A.A.S.
MIDI Music Production - A.A.S.

South Plains College
Levelland, TX 79336
www.southplainscollege.edu
Sound Technology - A.A.S.

Southwestern College
Chula Vista, CA 91910
www.swc.cc.ca.us
Commercial Music - A.S.

Valencia Community College
Orlando, FL 32802
www.valenciacc.edu
Music Production Technology - A.S. Music Production Technology

Bachelor's Degree Programs

American University
Washington, DC 20016-8058
kotsbue.physics.american.edu
Audio Technology - B.S.

Appalachian State University
Boone, NC 28608
www.music.appstate.edu/recording
Music Industry Studies--Recording Concentration - B.S.- Music Industry Studies

Ball State University
Muncie, IN 47306
www.bsu.edu
Music Engineering Technology - B.Mus.

Barton College
Wilson, NC 27893
www.barton.edu
Audio Recording Technology - B.S.

Belmont University
Nashville, TN 37212
www.belmont.edu
Audio Engineering Technology - B.S.
Qualification Awarded on Completion: B.S.

Berklee College of Music
Boston, MA 02215
www.berklee.edu
Music Production and Engineering, Music Synthesis Majors - B.Mus.

Bowling Green State University
Bowling Green, OH 43403
www.bgsu.edu
Recording Technology - RecordingTechnology Minor with any University Major

California Institute of the Arts
Valencia, CA 91355
music.calarts.edu
Multi-Focus Music Technologies Program - B.F.A.

California Lutheran University
Thousand Oaks, CA 91360
www.clunet.edu
B.A. in music with an emphasis in music technology

California State University Dominguez Hills
Carson, CA 90747
www.csudh.edu
Audio and Multimedia Production - B.A.
Audio Recording I and II - B.A.

California State University, Chico
Chico, CA 95929-0805
www.csuchico.edu
Recording Arts - B.A. in Music with option in Recording Arts (minor available)

Capital University Conservatory of Music
Columbus, OH 43209-2394
www.capital.edu
Music Technology - B.Mus.
Professional Studies - B.A., Emphasis in Music Technology

Clemson University
Clemson, SC 29634
www.clemson.edu
Audio Engineering - B.A.

Cleveland Institute of Music
Cleveland, OH 44106
www.cim.edu
Audio Recording - B.Mus.

Cogswell College
Sunnyvale, CA 94089
www.cogswell.edu
Digital Audio Technology - B.S.

Columbia College Chicago, Audio Arts and Acoustics Dept
Chicago, IL 60605
www.colum.edu
Acoustics, Sound Contracting Audio for the Visual Medium, Recording

Elmhurst College
Elmhurst, IL 60126
Audio Engineering – B.S. or B.Mus.
Production of Sound Recordings - B.S. or B.Mus.

Full Sail Real World Education
Winter Park, FL 32792
www.fullsail.com
Entertainment Business - B.S.

Georgia Southern University
Statesboro, GA 30460
mustech.mus.georgiasouthern.edu/
B.S. in Information Technology with a Second Discipline in Music Technology - B.S.

Greenville College
Greenville, IL 62246
www.greenville.edu
Contemporary Christian Music - B.S., with Recording and Production emphasis
Digital Media - B.S. with music emphasis

Hampton University
Hampton, VA 23668
Music Engineering Technology - B.S.

Indiana University School of Music
Bloomington, IN 47405
www.music.indiana.edu
Audio Recording - B.S.

Ithaca College School of Music
Ithaca, NY 14850
www.ithaca.edu/music
B.Mus. in Sound Recording Technology

Keene State College
Keene, NH 03435
www.keene.edu/programs/mu/
Music Technology - B.A.

Lebanon Valley College of Pennsylvania
Annville, PA 17003
Sound Recording Technology - B.Mus.
Qualification Awarded on Completion: B.M.

Mercy College
White Plains, NY 10601
www.mercy.edu/cda/
Music Industry & Technology Program - B.S.

Middle Tennessee State University
Murfreesboro, TN 37132
www.mtsu.edu
Recording Industry-Production and Technology - B.S.
The Recording Industry-Music Business - B.S.

Missouri Western State University
Saint Joseph, MO 64507
www.missouriwestern.edu/music
Music – B.S. in Music: Business/Music Technology
Bachelor of Science in Music - B.S. in Music: Business/Music Technology

The New England Institute of Art
Brookline, MA 02445
www.neia.artinstitutes.edu
Audio & Media Technology - B.S.

New England School of Communicatons
Bangor, ME 04401
www.nescom.edu
Audio Engineering - Bachelors Degree in Communications

New York University
New York, NY 10012
www.nyu.edu/education/music/mtech
Music Technology Program - B.Mus.

Oberlin Conservatory
Oberlin, OH 44074
Music and Technology - B.Mus.

Oral Roberts University
Tulsa, OK 74171
www.oru.edu
Music/Technology - B.A.
Mass Media/Multimedia Production - B.S.

Peabody Institute of The Johns Hopkins University
Baltimore, MD 21202
www.peabody.jhu.edu/
Recording Arts and Sciences - B.Mus.

Radford University
Radford, VA 24142
www.radford.edu/~cmt-web
Music and Technology Degree - B.Mus. (Music and Technology)

Sonic Arts Center @ The City College of New York
New York, NY 10031
sonic.arts.ccny.cuny.edu
Music and Audio Technology - BFA in Music (Music and Audio Technology)

State University of New York at Fredonia School of Music
Fredonia, NY 14063
www.fredonia.edu/som/srt/index.html

University of Colorado at Denver
Denver, CO 80217-3364
www.cudenver.edu
Music/Music Industry Studies - B.S.
Music/Recording Arts - B.S.

University of Hartford, College of Engineering
West Hartford, CT 06117
uhavax.hartford.edu/acoustics
Acoustics and Music Engineering - B.S. Engineering (BSE), B.S Mechanical Engineering, Optional EE minor
Mechanical Engineering with Acoustics Concentration - B.S. Engineering (BSE), B.S Mechanical Engineering, Optional EE minor

University of Hartford, The Hartt School of Music
West Hartford, CT 06117
uhaweb.hartford.edu/musicprod
Music Production and Technology - B.Mus.

University of Hartford, Ward College of Technology
West Hartford, CT 06117
uhaweb.hartford.edu/wardweb
Audio Engineering Technology - B.S.

University of Maine at Augusta
Augusta, ME 04330
Audio Recording (MUS 219)

University of Massachusetts Lowell
Lowell, MA 01854
www.uml.edu

Sound Recording Technology - B.Mus.
Sound Recording Technology for Computer Science majors - B.Sc. in CS (SRT minor)
Sound Recording Technology for Elec. Eng. Majors - Bachelor of Engineering (minor SRT)

University of Miami, Engineering
Coral Gables, FL 33124
Audio Engineering - BSEE

University of Miami, Music
Coral Gables, FL 33124
www.music.miami.edu
Music Engineering - B.Mus.

University of New Haven
West Haven, CT 06516
www.newhaven.edu
Music and Sound Recording - B.A., B.S.
Music Industry - B.A.

University of North Carolina at Asheville
Asheville, NC 28804-3299
Music with Recording Arts - B.S.

University of South Carolina
Columbia, SC 29208
Advanced Audio Recording Techniques - B.A. Media Arts
Aesthetics of Sound Imaging - B.A. Media Arts
Film Scoring - B.A. Media Arts
Introduction to Audio Recording Techniques - B.A. Media Arts
Use of Sound in Media Arts - B.A. Media Arts

Purdue University
West Lafayette, IN 47907
www.purdue.edu
Interdisciplinary Engineering - B.S.E. ; B.S.

Saint Mary's University of MN
Winona, MN 55987
www.smumn.edu
Music Recording and Technology - B.A.

School of Telecommunications at Ohio University
Athens, OH 45701
Audio Engineering - B.S.

Sonoma State University
Rohnert Park, CA 94928
www.sonoma.edu
Recording Arts Minor - B.A. in Music or Communication Studies with a minor concentration in recording arts

Southwest Texas State University
San Marcos, TX 78666
www.swt.edu
Sound Recording Technology - B.M.

University of Arkansas at Pine Bluff
Pine Bluff, AR 71601
www.uapb.edu
Music- B.S. with emphasis in Sound Recording Technology

University of Cincinnati - College Conservatory of Music
Cincinnati, OH 45221-0003
www.uc.edu
Theatre Design & Production - B.F.A.

University of Memphis
Memphis, TN 38152
music.memphis.edu
Recording Technology and Music Business - B.Mus.

University of Michigan, School of Music
Ann Arbor, MI 48109-2085
www.music.umich.edu
Sound Engineering - B.S.

University of Oregon
Eugene, OR 97403-1225
music.uoregon.edu
Music, Technology Option - B.S.

University of Southern California
Los Angeles, CA 90089-0851
Music Industry - B.S.
Music Recording - B.S. , B.Mus.

University of Southwestern Louisiana School of Music
Lafayette, LA 70504
music.louisiana.edu
Music Media - B.Mus., B.S., B.A.

University of Wisconsin-Oshkosh
Oshkosh, WI 54901
Bachelor of Music with emphasis in Music Merchandising - B.Mus.
Bachelor of Music with emphasis in Recording Technology - B.Mus.

Webster University
Webster Groves, MO 63119
www.webster.edu
Audio Production - B.A.

Master's Degree Programs

American University
Washington, DC 20016-8058
kotsbue.physics.american.edu
Interdisciplinary Studies in Audio - M.S.

Georgia Southern University
Statesboro, GA 30460
mustech.mus.georgiasouthern.edu
M.Mus. in Music Technology

McGill University, Faculty of Music
Montreal, QC H3A 1E3, Canada
www.music.mcgill.ca
Sound Recording - M.Mus.

Middle Tennessee State University
Murfreesboro, TN 37132
www.mtsu.edu
Recording Arts & Technologies - M.F.A.

New York University
New York, NY 10012
www.nyu.edu
Music Technology Program - M.Mus.

Peabody Institute of The Johns Hopkins University
Baltimore, MD 21202
www.peabody.jhu.edu
Audio Recording and Acoustics - M.A.

Penn State University Graduate Program in Acoustics
State College, PA 16804
www.acs.psu.edu
Continuing and Distance Education Program in Acoustics - M. Eng. in Acoustics

Purdue University
West Lafayette, IN 47907
www.purdue.edu
Theatre Audio Technology - M.F.A.
Theatre Sound Design - M.F.A.

University of California at Santa Barbara, Media Arts and Technology
Santa Barbara, CA 93106
www.mat.ucsb.edu
Electronic Music and Sound Design - M.A.
Multimedia Engineering - M.S.

University of Cincinnati - College Conservatory of Music
Cincinnati, OH 45221-0003
www.uc.edu
Theatre Sound Design - M.F.A.

University of Colorado at Denver
Denver, CO 80217-3364
www.cudenver.edu
Recording Arts - M.S.

University of Massachusetts Lowell
Lowell, MA 01854
www.uml.edu
Sound Recording Technology - M.Mus.

University of Miami, Music
Coral Gables, FL 33124
www.music.miami.edu
Music Engineering - M.S.

University of Missouri & Kansas City, Dept. of Theatre
Kansas City, MO 64110
Professional Sound Design, A-D - M.F.A.

University of Oregon
Eugene, OR 97403-1225
music.uoregon.edu
Intermedia Music Technology - M.Mus.

Index

A

B

E

F

G

H

I

J

K

L

M

R

S

T

U

V

W

About the Author

Celeste Baine is an avid drummer, a biomedical engineer, the director of the Engineering Education Service Center and the award-winning author of *Is There an Engineer Inside You: A Comprehensive Guide to Career Decisions in Engineering, The Fantastical Engineer: A Thrillseeker's Guide to Careers in Theme Park Engineering, High Tech Hot Shots: Careers in Sports Engineering* and *Teaching Engineering Made Easy: A Friendly Introduction to Engineering Activities for Middle School Teachers.* She is the author of the engineering section of the World Book Encyclopedia's Science Year and the principal trainer for the Academy of Engineering LEGO laboratory. She won the 2005 Norm Augustine Award for Engineering Communications which is given to an engineer who has demonstrated the capacity for communicating the excitement and wonder of engineering; the 2004 American Society for Engineering Education's Engineering Dean Council's Award for the Promotion of Engineering Education and Careers; and, she is also listed on the National Engineers Week Web site as one of 50 engineers you should meet!

Other Engineering Career Publications by Celeste Baine

The Fantastical Engineer: A Thrillseeker's Guide to Careers in Theme Park Engineering. (May 2007)

High Tech Hot Shots: Careers in Sports Engineering. $19.95

Is There an Engineer Inside You? A Comprehensive Guide to Career Decisions in Engineering (Second Edition). $17.95

Is There a Civil Engineer Inside You? A Student's Guide to Exploring Careers in Civil Engineering & Civil Engineering Technology. $7.95

Is There a Computer Engineer Inside You? A Student's Guide to Exploring Careers in Computer Engineering & Computer Engineering Technology. $7.95

Is There a Mechanical Engineer Inside You? A Student's Guide to Exploring Careers in Mechanical Engineering & Mechanical Engineering Technology. $7.95

Is There a Chemical Engineer Inside You? A Student's Guide to Exploring Careers in Chemical Engineering. $7.95

Is There a Biomedical Engineer Inside You? A Student's Guide to Exploring Careers in Biomedical Engineering & Biomedical Engineering Technology. $7.95

Is There an Electrical Engineer Inside You? A Student's Guide to Exploring Careers in Electrical Engineering & Electrical Engineering Technology. $7.95

Is There a Manufacturing Engineer Inside You? A Student's Guide to Exploring Careers in Manufacturing Engineering & Manufacturing Engineering Technology. $7.95

Engineering Career DVDs!

Engineers Can Do Anything!

20 minutes, $24.95, Pub Date: May 2005

Anyone who is interested in becoming an engineer should watch this DVD because it will open doors and spark imagination. See engineering in a new and fascinating light from competition battlefields to summer camps to intriguing new career possibilities. Infused with the music of Geek Rhythms and narrated by award winning author Celeste Baine, this DVD is sure to deliver the facts about engineering careers in a fun, motivating and engaging way. Engineers Can Do Anything! will change the way you see engineering and show you how engineers are involved in every segment of society.

"This high-energy DVD is one of the best new resources I've seen for exciting and motivating the next generation of engineers. It is great for students, teachers, classrooms, summer programs, parents, and anyone who knows a young person with a dream to change the world! No matter what your background, this production will open your eyes to looking at engineering in new ways . Show this DVD to every student group you can!"

-Peter Larson, Youth Programs Director, Michigan Tech University

The Road Ahead Choosing an Engineering School!

12 minutes, $14.95, Pub Date: May 2006

Which engineering school is right for you? An excellent first step, The Road Ahead motivates and helps students begin to sift through the maze of choices available when considering colleges. The Road Ahead answers the questions that help every student figure out what they want and what they need from a college. Taken to heart, this video will help you choose a school in less time, with less struggle and less uncertainty.